Lecture Notes in Mathematics 　2083

Editors:
J.-M. Morel, Cachan
B. Teissier, Paris

For further volumes:
http://www.springer.com/series/304

Anna M. Bigatti • Philippe Gimenez
Eduardo Sáenz-de-Cabezón

Editors

Monomial Ideals, Computations and Applications

Editors
Anna M. Bigatti
Dipartimento di Matematica
Universitá degli Studi di Genova
Genova, Italy

Philippe Gimenez
Dpto. de Álgebra, Análisis Matemático,
 Geometría y Topología
University of Valladolid
Valladolid, Spain

Eduardo Sáenz-de-Cabezón
Matemáticas y Computación
University of La Rioja
Logroño, Spain

ISBN 978-3-642-38741-8   ISBN 978-3-642-38742-5 (eBook)
DOI 10.1007/978-3-642-38742-5
Springer Heidelberg New York Dordrecht London

Lecture Notes in Mathematics ISSN print edition: 0075-8434
                               ISSN electronic edition: 1617-9692

Library of Congress Control Number: 2013945178

Mathematics Subject Classification (2010): 13-02, 13C15, 13D45, 13F55

Printed on acid-free paper

Springer is part of Springer Science+Business Media (www.springer.com)

Contents

Part II   Edge Ideals

Part III   Local Cohomology

Introduction

Monomial ideals and algebras are among the simplest structures in commutative algebra and the main objects in combinatorial commutative algebra. One can highlight several characteristics of these objects that make them of special interest.

One of the main points related to monomial ideals in commutative algebra and algebraic geometry is that some problems on general polynomial ideals can be reduced to the special case of monomial ideals through the theory of Gröbner basis. For example, the dimension of an affine (or a projective) algebraic variety defined by an arbitrary polynomial ideal I coincides with the dimension of the variety defined by a monomial ideal associated with I, the *initial ideal* (or *leading term ideal*) of I with respect to a term ordering. On the other hand, the dimension of the variety defined by a monomial ideal can be read off from its generators easily. This example illustrates how a problem in commutative algebra and algebraic geometry (the computation of the dimension of an arbitrary algebraic variety) can be reduced to an easier problem on monomial ideals. From a computational point of view, the combinatorial nature of monomial ideals makes them particularly well suited for the development of algorithms to work with and then generate algorithms focused on more general structures. The most important computer algebra systems specialized in polynomial computations, like CoCoA-5 [39], *Macaulay 2* [47] or Singular [80] which are, nowadays, a fundamental tool for the working commutative algebraist or algebraic geometer, make a heavy use of this fact. Besides its use for computational issues, the theory of Gröbner basis turned out to be a powerful tool for solving theoretical problems in commutative algebra and algebraic geometry.

Another fundamental aspect of monomial ideals is that their combinatorial structure connects them to other combinatorial objects and allows the resolution of problems at each side of this correspondence with the techniques of each of the respective areas. The work of Melvin Hochster and Richard Stanley, in particular [180], gave an important impetus to the development of these ideas, and the classical book of Stanley, [182], reflects how the areas of combinatorics and commutative algebra clearly interact. This led to the creation of a new and very active field, *combinatorial commutative algebra*.

The last decade has witnessed a plethora of different ideas, results and approaches to monomial ideals and their role in commutative algebra and its applications. This volume presents three different and representative topics on this area. They are written by leading experts in the field and give an account of the state of the art of the work on monomial ideals. The topics and the way to present them have been chosen so that the peculiarities of research in monomial ideals are highlighted together with the richness of approaches that these structures allow. The order of the chapters intend to guide the reader in the discovery of the variety of topics that arise in research on monomial ideals. The starting point of the book is in commutative algebra, the natural context in which monomial ideals are found. In the first part of this monograph, we can see how we can take advantage of the combinatorial nature of this type of ideals and use powerful tools of commutative algebra. From this conjunction, one obtains a first glance on the particular techniques one can use working with monomial ideals. The second part of the book shows how monomial ideals techniques share a common nature with certain combinatorial objects and tools. In this common area where monomial ideals lie, there are algebraic techniques that can be used for combinatorial problems and vice versa. Finally, in the third part one can see that the constructive character of the tools used when working with monomial ideals can also be used in topics of a more general and abstract nature in commutative algebra. After the journey through the three parts of this volume, the reader can have a neat impression on the essence of the work with monomial ideals by studying up-to-date research on three different contexts. The fact that each chapter includes a computer tutorial contributes to highlight one of the peculiarities of monomials ideals, that they are particularly well suited for producing algorithms and data types to study algebraic and combinatorial topics.

We now discuss the topics and authors of the three parts of this volume.

The first chapter of the book, by Jürgen Herzog, gives a survey on Stanley decompositions and their relation to the depth of a module and discusses the conjecture by Richard Stanley regarding this relationship. The conjecture remains wide open and research around its solution has provided a collection of concepts and results that involved many different authors. Jürgen Herzog has been very actively involved in this line of research and he gives in this chapter a first-hand survey where the conjecture is discussed in all its different aspects. The tutorial of this part, written by Anna M. Bigatti and Emanuela De Negri in the second chapter, uses the computer algebra system CoCoA-5 [39] to compute Stanley decompositions and allows the reader to explore the computational aspects of this topic.

The third chapter, by Adam Van Tuyl, features the basic properties of edge and cover ideals and introduces some current research themes. This topic has recently received much attention, producing interesting connections and beautiful results in the intersection of combinatorics and commutative algebra. The tutorial of this part, also written by Adam Van Tuyl in the fourth chapter, includes an introduction to the package EdgeIdeals [69] distributed with the computer algebra system *Macaulay 2* [47]. It also includes a list of exercises that can be solved, most of them, using the package EdgeIdeals and that illustrate the concepts and results introduced in this part.

Finally, the last part of the book gives an overview of some results that have been developed in recent years about the structure of local cohomology modules supported on a monomial ideal. This part is written by Josep Àlvarez Montaner in the fifth chapter where the interplay of multi-graded commutative algebra, combinatorics and D-modules theory is highlighted, providing different points of view on this subject that led to fundamental improvements in our understanding of these notions. The tutorial accompanying this part, written by Josep Àlvarez Montaner and Oscar Fernández-Ramos in the last chapter, proposes several examples and exercises to perform computations on local cohomology modules supported on monomial ideals. As for the second part of this volume, this tutorial uses the computer algebra system *Macaulay 2* [47], and the package EdgeIdeals, already featured in the fourth chapter, is used in some of the exercises.

This book originated from a series of lectures given by the authors at the conference *MONICA: MONomial Ideals, Computations and Applications* held at CIEM, Castro Urdiales (Cantabria, Spain) in July 2011. The MONICA meeting was organized by the editors of this volume and included, together with the three lecture series and their tutorials, some contributed talks by the participants, with a wide range of topics around monomial ideals and algebras. We want to thank all the participants and, in particular, the invited speakers and computer tutorial authors for the high quality of their contributions, showing the variety and excellence of research on monomial ideals and algebras. Last but not least, we finally want to thank i-math, CIEM, the city of Castro Urdiales, the University of Valladolid and the Spanish government (*Ministerio de Ciencia e Innovación*, grants MTM2010-20279-C02-02 and MTM2009-13842-C02-01) for providing financial support for the MONICA conference. We hope that this volume contributes to cover the lack of survey bibliography in an area in which such intense research is done in many directions.

Genova, Italy

Valladolid, Spain

Logroño, Spain

Anna Maria Bigatti

Philippe Gimenez

Eduardo Sáenz-de-Cabezón

Contributors

Anna Maria Bigatti Dipartimento di Matematica, Università degli Studi di Genova, Genova, Italy

Emanuela De Negri Dipartimento di Matematica, Università degli Studi di Genova, Genova, Italy

Oscar Fernández-Ramos Dipartimento di Matematica, Università degli Studi di Genova, Genova, Italy

Jürgen Herzog Universität Duisburg-Essen, Fachbereich Mathematik, Essen, Germany

Josep Àlvarez Montaner Departamento de Matemàtica Aplicada I, Universitat Politècnica de Catalunya, Barcelona, Spain

Adam Van Tuyl Department of Mathematical Sciences, Lakehead University, Thunder Bay, ON, Canada

Part I
Stanley Decompositions

A Survey on Stanley Depth

Jürgen Herzog

Introduction

At the MONICA conference "MONomial Ideals, Computations and Applications" at the CIEM, Castro Urdiales (Cantabria, Spain) in July 2011, I gave three lectures covering different topics of Combinatorial Commutative Algebra: (1) A survey on Stanley decompositions. (2) Generalized Hibi rings and Hibi ideals. (3) Ideals generated by two-minors with applications to Algebraic Statistics. In this article I will restrict myself to give an extended presentation of the first lecture. The CoCoA tutorials following this survey will deal also with topics related to the other two lectures. Complementing the tutorials, the reader finds in [165] a CoCoA routine to compute the Stanley depth for modules of the form I/J, where $J \subset I$ are monomial ideals.

In his famous article "Linear Diophantine equations and local cohomology", Richard Stanley [181] made a striking conjecture predicting an upper bound for the depth of a multigraded module. This conjectured upper bound is nowadays called the Stanley depth of a module. The Stanley depth is of a rather combinatorial nature while the depth is a homological invariant. The definition of Stanley depth is given in Sect. 1. The conjecture is on so far striking as it compares two invariants of modules of very different nature. At a first glance it seems to be no relation among these two invariants. The conjecture was made in 1982, and to best of my knowledge it is Apel who first studied this conjecture intensively and proved it in some special cases in his papers [12, 13]. His papers appeared in 2003. Stanley decompositions were then considered again in 2006 in my joint paper [95] with Popescu and since then the topic has become very popular with numerous publications regarding different

J. Herzog (✉)
Universität Duisburg-Essen Fachbereich Mathematik, Campus Essen 45117 Essen, Germany
e-mail: juergen.herzog@uni-essen.de

A.M. Bigatti et al. (eds.), *Monomial Ideals, Computations and Applications*,
Lecture Notes in Mathematics 2083, DOI 10.1007/978-3-642-38742-5_1,
© Springer-Verlag Berlin Heidelberg 2013

aspects of Stanley depth. Though there has been a lot of efforts to prove or disprove Stanley's conjecture, it is still widely open.

I tried to give a rather comprehensive list of references to papers dealing with Stanley depth. It is impossible to describe and to discuss the content of all these papers in this survey. In the very last section of this survey however I try to list some of the most remarkable results which are not discussed in these notes to give the reader an orientation on what is known, and I also list a few open questions which I think would be interesting to deal with.

This survey on Stanley depth naturally describes the theory from a viewpoint based on my own and my coauthors contributions, as presented in the papers [95, 98, 99, 102]. In Sect. 1 we recall Stanley's conjecture in its original form and fix the terminology and notation used throughout this notes. In Sect. 2 we introduce the concept of prime filtrations and show how they induce Stanley decompositions, thereby proving that Stanley decompositions actually exist. The discussions there lead to an invariant which is denoted by fdepth (filter depth). Then in the following Sect. 3 we show that the invariant fdepth gives a lower bound for the Stanley depth and we also present an upper bound that was proved by Apel. This upper bound implies in particular that the Stanley depth is bounded above by the Krull dimension.

Section 4 is devoted to the theory initiated by Dress [50] who showed that a simplicial complex is shellable (in the non-pure sense) if and only if its Stanley–Reisner ring is clean, meaning that it has a prime filtration with all its prime factors being minimal prime ideals. Here we present the proof of the fact that the Stanley–Reisner ring of a shellable simplicial complex is clean. In the context of Stanley decompositions this result is remarkable as it implies that the Stanley–Reisner ring of a shellable simplicial complex satisfies Stanley's conjecture. To extend this result to K-algebras with monomial relations, not necessarily squarefree, pretty clean filtrations were introduced in [95]. Modules with pretty clean filtrations also satisfy Stanley's conjecture. A pretty clean module which has no embedded prime ideals is actually clean. At the end of Sect. 4 we show that a polynomial ring modulo a Borel type ideal (ideal of nested type) is always pretty clean.

Long before Stanley made his conjecture, Janet [115] in 1924 gave an explicit algorithm to produce a Stanley decomposition of a monomial ideal. In Sect. 5 we describe his algorithm. Janet decompositions from the viewpoint of Stanley depth are not optimal. They rarely give Stanley decompositions providing the Stanley depth of a monomial ideal. However one obtains the result that the Stanley depth of a monomial ideal is at least 1.

A well-known result in Commutative Algebra is the fact that the depth of a module after reduction by a non-zero divisor drops exactly by one. In Sect. 6 we present the result of Rauf which says that the Stanley depth drops at least by one after reduction by a variable which is a non-zero divisor on the module. This result has two nice consequences which we cite from [32]. The first is that Stanley depth zero implies depth zero, the second is that the kth syzygy module of a $\mathbb{Z}^n$-graded module has Stanley depth at least k. The second result (with a more difficult argument) was first shown in [62].

To compute the Stanley depth of a module one has in principle to consider all possible Stanley decompositions. These are infinitely many. So the question arises whether the Stanley depth of a module can actually be computed. Indeed, up to date, no algorithm for computing the Stanley depth of a multigraded module is known. Only the Stanley depth of I/J for monomial ideals $J \subset I$ can be computed. This is the content of Sect. 7. Here we reproduce the theory as given in [101]. A slightly revised version of the paper appeared later in Journal of Algebra where Miller's functors as defined in [140] are used. The method of computing the Stanley depth of I/J consists in attaching to I/J a finite poset $P^g_{I/J}$, called the characteristic poset of I/J, with the property that each of its interval partitions yields a Stanley decomposition of I/J. The remarkable fact is that among the Stanley decompositions obtained in this way there is one which gives the Stanley depth. So in principle, the Stanley depth of I/J can be a computed in a finite number of steps. From a computational point of view however the method is not very efficient. Rinaldo in [165] uses this method, eliminating unnecessary interval partitions, for a routine to compute the Stanley depth [165]. Apart from the computational aspects, this approach via interval partitions has also some interesting theoretical aspects. For example it can be shown, as outlined in Sect. 7, that Stanley's conjecture on Stanley decompositions implies another one of Stanley's conjectures which asserts that each Cohen–Macaulay simplicial complex is partionable.

The characteristic poset of I/J can be used to define the skeletons of I/J, which when specialized to the Stanley–Reisner ideal of a simplicial complex gives the Stanley–Reisner ideals of its skeletons. This concept is discussed in Sect. 8 following the article [99]. It is used to determine the depth of S/I for a monomial ideal $I \subset S$, generalizing a corresponding result of Hibi in [104]. Moreover, this concept of skeletons is used to show that Stanley's conjecture holds for all K-algebras with monomial relations if and only if it holds for all Cohen–Macaulay K-algebras with monomial relations.

The method described in Sect. 7 to compute the Stanley depth leads to interesting combinatorial considerations in even seemingly simple cases as for example in the case of the graded maximal ideal m of a polynomial ring in n variables. While the depth of m is easily seen to be 1, Biró et al. showed in [18] that the Stanley depth of m is $\lceil n/2 \rceil$. We present in Sect. 9 the main ideas of their nice combinatorial arguments. The result shows that the Stanley depth of module may exceed its depth by any amount. The combinatorial techniques developed in [18] have been extended and refined later by many authors. Here we use these extensions to show, following the arguments of [62], that the lower bound for the Stanley depth of a squarefree monomial ideal in a polynomial ring in n variables is approximately of size $\sqrt{n}$.

In Sect. 10 we describe the relationship between squarefree Stanley decompositions of the Stanley–Reisner ring of a simplicial complex and its Alexander dual, as it is done in [179]. This considerations naturally suggest to define a new invariant for $\mathbb{Z}^n$-graded modules denoted sreg (Stanley regularity). The Stanley regularity of a Stanley–Reisner ideal of a simplicial complex on the vertex set $[n]$ and the Stanley depth of the Stanley–Reisner ring of its Alexander dual always add up to n. This is a formula analogue to that discovered by Terai relating depth and regularity.

The conjecture of Soleyman Jahan, in some sense dual to that of Stanley, says that the Stanley regularity of a $\mathbb{Z}^n$-graded S-module is always less than or equal to its regularity.

In [129], Lyubeznik gave a nice lower bound for the depth of S/I where I is a monomial ideal, called the size of I. Provided the primary decomposition of I is known, the size of I is an easy to compute invariant of S/I which does not depend on the characteristic of the base field. The main purpose of Sect. 11 is to show that the same lower bound holds for the Stanley depth which, assuming that Stanley's conjecture holds true, is not surprising. Considering Alexander duality one is guided to define the so-called cosize of an ideal, and obtains that both the regularity as well as the Stanley regularity are bounded above by the cosize. This section reflects the main results of the paper [97].

A $\mathbb{Z}^n$-graded S-module, S a polynomial ring, may naturally be considered a $\mathbb{Z}$-graded module over S, where S is given the standard grading. Then the Hilbert series of M viewed as a $\mathbb{Z}$-graded module can be computed from each of its Stanley decompositions. This leads to an upper bound for the Stanley depth which only depends on the Hilbert series of the module. This upper bound is called the Hilbert depth, introduced by Bruns and his coauthors in [31]. In Sect. 12 we outline the idea of this concept. Amazingly, the Hilbert depth always exceeds the depth of a module, as shown in [31]. A different proof of this fact is given here for S/I where I is a monomial ideal. One would expect that it is easy to compute the Hilbert depth of a module, once its Hilbert function is known. But it turns out that even for the powers of the maximal ideal, the computation of the Hilbert depth leads to difficult numerical computations. Some of the results by Bruns and his group on the Hilbert depth is quoted in the last section, where also a few more nice results, not discussed in this survey, are mentioned.

Of course the most challenging open problem is to prove or to disprove Stanley's conjecture. To find a counter example would be quite hard. By what has been shown in this survey one would find a possible counter example among the Stanley–Reisner rings of a Cohen–Macaulay simplicial complex which is not shellable. On the other hand, there are many other attractive problems related to Stanley depth, not directly related to Stanley's conjecture. Some of them are listed in the last section.

1 Basic Definitions and Concepts

Richard Stanley [181] in his article "Linear Diophantine equations and local cohomology", made the following striking conjecture concerning the depth of multigraded modules.

Conjecture 1 (Stanley). Let R be a finitely-generated $\mathbb{N}^n$-graded K-algebra (where $R_0 = K$ as usual), and let M be a finitely generated $\mathbb{Z}^n$-graded R-module. Then there exist finitely many subalgebras $S_1, \ldots, S_t$ of R, each generated by algebraically independent $\mathbb{N}^n$-homogeneous elements of R, and there exist $\mathbb{Z}^n$-homogeneous elements $\eta_1, \ldots, \eta_t$ of M, such that

$$M = \bigoplus_{i=1}^{t} \eta_i S_i, \quad \text{(vector space direct sum)}$$

where $\dim(S_i) \geq \operatorname{depth}(M)$ for all i, and where $\eta_i S_i$ is a free S_i-module (of rank one). Moreover, if K is infinite and under a given specialization to an $\mathbb{N}$-grading R is generated by R_1, then we can choose the ($\mathbb{N}^n$-homogeneous) generators of each S_i to lie in R_1.

Stanley calls the above direct sum decomposition of M a combinatorial decomposition of M. Nowadays such a decomposition is called a Stanley decomposition.

In the same article Stanley proved this conjecture for decompositions of monoid rings, see [182, Theorem 5.2].

In this survey we concentrate our attention to the case that M is a finitely generated $\mathbb{Z}^n$-graded S-module, where $S = K[x_1, \ldots, x_n]$ is the polynomial ring over a field K.

Definition 2. Let M be a $\mathbb{Z}^n$-graded S-module. Let $m \in M$ be homogeneous, and $Z \subset X = \{x_1, \ldots, x_n\}$. Then the $K[Z]$-submodule $mK[Z]$ of M is called a *Stanley space* of M, if $mK[Z]$ is a free $K[Z]$-submodule of M, and $|Z|$ is called the *dimension* of $mK[Z]$.

A *Stanley decomposition* $\mathscr{D}$ of M is a decomposition of M as a direct sum of $\mathbb{Z}^n$-graded K-vector space

$$\mathscr{D}: M = \bigoplus_{j=1}^{r} m_j K[Z_j],$$

where each $m_j K[Z_j]$ is a Stanley space of M.

We set

$$\operatorname{sdepth}(\mathscr{D}) = \min\{|Z_j| : j = 1, \ldots, r\},$$

and

$$\operatorname{sdepth}(M) = \max\{\operatorname{sdepth}(\mathscr{D}) : \mathscr{D} \text{ is a Stanley decomposition of } M\}.$$

The number $\operatorname{sdepth}(M)$ is called the *Stanley depth* of M.

With the notation introduced, Stanley's conjecture can be phrased as follows:

Conjecture 3 (Stanley). $\operatorname{sdepth}(M) \geq \operatorname{depth}(M)$.

The conjecture is widely open. Nevertheless there have been interesting developments in recent years in the context of this conjecture.

Fig. 1 A Stanley
decomposition of I and S/I

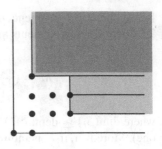

Of particular interest is the case when M is isomorphic to a monomial ideal $I \subset S$ or isomorphic to the residue class ring S/I of a monomial ideal I. The monomials $u \in I$ form a homogeneous K-basis of I, while the residue classes modulo I of the monomials $u \in S \setminus I$ form a homogeneous basis of S/I. Therefore, we often identify S/I with the $\mathbb{Z}^n$-graded K-subvector space I^c of S which is generated by all monomials $u \in S \setminus I$.

Figure 1 displays Stanley decompositions of $I = (x_1^3 x_2, x_1 x_2^3)$ and S/I, namely

$$\mathscr{D}_1 : I = x_1 x_2^3 K[x_1, x_2] \oplus x_1^3 x_2^2 K[x_1] \oplus x_1^3 x_2 K[x_1],$$

and

$$\mathscr{D}_2 : S/I = K[x_2] \oplus x_1 K[x_1] \oplus x_1 x_2 K \oplus x_1 x_2^2 K \oplus x_1^2 x_2 K \oplus x_1^2 x_2^2 K.$$

We have $\mathrm{sdepth}(I) \geq \mathrm{sdepth}(\mathscr{D}_1) = 1 = \mathrm{depth}(I)$, and $\mathrm{sdepth}(S/I) \geq \mathrm{sdepth}(\mathscr{D}_2) = 0 = \mathrm{depth}(S/I)$. Thus Stanley's conjecture holds in this particular case.

The natural question arises whether there exists always a Stanley decomposition (otherwise Definition 2 wouldn't make any sense), and whether we can compute the Stanley depth.

Note that whenever M contains a Stanley space $mK[Z]$ of dimension > 0, then M admits infinitely many Stanley decompositions. Indeed, suppose that a Stanley decomposition $\mathscr{D}$ contains the Stanley space $mK[Z]$ with $|Z| > 0$ as a summand. Then, among many other possibilities, we may replace the summand $mK[Z]$ for any r and any k with $x_k \in Z$ by the direct sum

$$\bigoplus_{i=0}^{r-1} x_k^i mK[Z'] \oplus x_k^r mK[Z],$$

where $Z' = Z \setminus \{x_k\}$.

Thus, since there are in general infinitely many different Stanley decompositions of a module, it is not at all clear how to compute the Stanley depth. We will deal with this problem in Sect. 7. In the next section we will show that any $\mathbb{Z}^n$-graded module admits indeed a Stanley decomposition.

2 Prime Filtrations and Stanley Decompositions

Let $S = K[x_1, \ldots, x_n]$ be the polynomial ring in n variables over a field K and M a finitely generated $\mathbb{Z}^n$-graded S-module. It is known that the associated prime ideals of M are monomial ideals, and that any monomial prime ideal is of the form $P_F = (x_i : i \notin F)$ for some $F \subset [n]$.

Definition 4. A chain

$$\mathscr{F} : 0 = M_0 \subset M_1 \subset \cdots \subset M_r = M$$

of $\mathbb{Z}^n$-graded submodules of M such that $M_i/M_{i-1} \cong (S/P_i)(-a_i)$ with P_i a monomial prime ideal and $a_i \in \mathbb{Z}^n$ is called a *prime filtration* of M. The *support* of the prime filtration $\mathscr{F}$ is the set $\{P_1, \ldots, P_r\}$, denoted $\mathrm{Supp}(\mathscr{F})$.

Prime filtrations always exist: Let $P_1 \in \mathrm{Ass}(M)$. Since M is $\mathbb{Z}^n$-graded, P_1 is a monomial prime ideal and $P_1 = \mathrm{Ann}(m_1)$ where m_1 is a homogeneous element of M. Set $M_1 = Sm_1$. Then $M_1 \cong (S/P_1)(-a_1)$ where $a_1 = \deg m_1$. If $M_1 = M$, the assertion follows. Otherwise $\mathrm{Ass}(M/M_1) \neq \emptyset$. Let $P_2 \in \mathrm{Ass}(M/M_1)$. Then there exists a $\mathbb{Z}^n$-graded submodule $M_2 \subset M$ with $M_1 \subset M_2$ such that $M_2/M_1 \cong (S/P_2)(-a_2)$ for some $a_2 \in \mathbb{Z}^n$. If $M_2 = M$, then $0 = M_0 \subset M_1 \subset M_2 = M$ is the desired prime filtration. If $M_2 \neq M$, we find in the same way as described before $M_3 \subset M$ with $M_3/M_2 \cong (S/P_3)(-a_3)$ where P_3 is a monomial prime ideal and $a_3 \in \mathbb{Z}^n$. Since M is Noetherian, this process must stop and hence yields the desired prime filtration.

This elementary fact from commutative algebra yield the desired existence theorem.

Theorem 5. *Let M be a finitely generated $\mathbb{Z}^n$-graded S-module. Then M admits a Stanley decomposition. More precisely, let*

$$\mathscr{F} : 0 = M_0 \subset M_1 \subset \cdots \subset M_r = M$$

be a prime filtration of M with $M_i/M_{i-1} \cong (S/P_i)(-a_i)$ and $M_i = M_{i-1} + Sm_i$ with m_i homogeneous of degree a_i. Then one obtains the Stanley decomposition

$$\mathscr{D}(\mathscr{F}) : M = \bigoplus_{i=1}^{r} m_i K[Z_i],$$

where $Z_i = \{x_j : x_j \notin P_i\}$.

The proof of the theorem is immediate: observe that $M \cong \bigoplus_{i=1}^{r} M_i/M_{i-1}$ as a direct sum of $\mathbb{Z}^n$-graded K-vector spaces, and that $S/P_i \cong K[Z_i]$ for all i.

Not each Stanley decomposition of M is of the form $\mathscr{D}(\mathscr{F})$ for a suitable prime filtration $\mathscr{F}$ of M. Indeed, a prime filtration $\mathscr{F}$ of M is essentially given by a sequence of homogeneous elements $m_1, \ldots, m_r$ of M with the property that

$M = Sm_1 + \cdots + Sm_r$ and that each of the colon ideals $(m_1, \ldots, m_{i-1}) : m_i$ is generated by a subset of the variables (in which case we say that the sequence $m_1, \ldots, m_r$ has linear quotients). Thus we see that a Stanley decomposition $\mathscr{D} : M = \bigoplus_{i=1}^{r} m_i K[Z_i]$ is induced by a prime filtration of M if and only if, after a suitable renumbering of the direct summands, we have that

$$(m_1, \ldots, m_{i-1}) : m_i = (x_j : x_j \notin Z_i) \quad \text{for all} \quad i.$$

Example 6. Consider the following Stanley decomposition

$$\mathscr{D} : (x_1, x_2, x_3) = x_1 x_2 x_3 K[x_1, x_2, x_3] \oplus x_1 K[x_1, x_2] \oplus x_2 K[x_2, x_3] \oplus x_3 K[x_1, x_3].$$

of the maximal ideal $(x_1, x_2, x_3) \subset K[x_1, x_2 x_3]$.

It is easily seen that no order of the elements $x_1 x_2 x_3, x_1, x_2, x_3$ forms a sequence with linear quotients. Thus this Stanley decomposition does not arise from a prime filtration.

The following characterization of Stanley decompositions arising from prime filtrations is taken from [101].

Proposition 7. *Let M be a finitely generated $\mathbb{Z}^n$-graded S-module with $\dim_K M_a \leq 1$ for all $a \in \mathbb{Z}^n$, and let $\mathscr{D} : M = \bigoplus_{i=1}^{r} u_i K[Z_i]$ be a Stanley decomposition of M. Then the following conditions are equivalent:*
(a) *$\mathscr{D}$ is induced by a prime filtration.*
(b) *After a suitable relabeling of the summands in $\mathscr{D}$, for $j = 1, \ldots, r$ the direct sum $M_j = \bigoplus_{i=1}^{j} m_i K[Z_i]$ is a $\mathbb{Z}^n$-graded submodule of M.*

Proof. (a) $\Rightarrow$ (b) follows immediately from the construction of a Stanley decomposition which is induced by a prime filtration.

(b) $\Rightarrow$ (a): We claim that $\mathscr{F} : 0 \subset M_1 \subset M_2 \subset \ldots \subset M_r = M$ is a prime filtration of M. First notice that for each j, the module M_j / M_{j-1} is a cyclic module generated by the residue class $\bar{m}_j = m_j + M_{j-1}$. Indeed, each element $m \in M_j$ can be written as $m = \sum_{k=1}^{j} m_k f_k$ with $f_k \in K[Z_k]$ for $k = 1, \ldots, j$. Therefore $\bar{m} = \bar{m}_j f_j$.

Next we claim that the annihilator of $\bar{m}_j$ is equal to the monomial prime ideal P generated by the variables $x_k \notin Z_j$. Since $\dim_K M_a \leq 1$ for all a, it follows that if $m \in M$ is homogeneous, then there is a unique Stanley space $m_i K[Z_i]$ such that $m \in m_i K[Z_i]$. Thus, if $x_k \notin Z_j$, then $x_k m_j \notin m_j K[Z_j]$ since $\deg x_k m_j \neq \deg m_j v$ for all monomials $v \in K[Z_j]$. Therefore, $x_k m_j \in m_i K[Z_i]$ for some $i < j$. This implies that $x_k \bar{m}_j = 0$ and shows that P is contained in the annihilator of $\bar{m}_j$. On the other hand, if v is a monomial in $S \setminus P$, then $v \in K[Z_j]$, and so $m_j v$ is a nonzero element in $m_j K[Z_j]$. This implies that v does not belong to the annihilator of $\bar{m}_j$ and shows that P is precisely the annihilator of $\bar{m}_j$. From all this we conclude that $\mathscr{D}$ is induced by $\mathscr{F}$.

Definition 8. Let M be a finitely generated $\mathbb{Z}^n$-graded S-module. The number

$$\text{fdepth}(M) = \min\{\text{sdepth}(\mathscr{D}(\mathscr{F})): \mathscr{F} \text{ is a prime filtration of } M\}$$

is called the *filtration depth* of M.

It is clear that $\text{fdepth}(M) \leq \text{sdepth}(M)$. Example 6 shows that this inequality may be strict. Indeed in that example we have $\text{fdepth}(x_1, x_2, x_3) = 1$ and $\text{sdepth}(x_1, x_2, x_3) = 2$.

We conclude this section by observing that for any prime filtration $\mathscr{F}$ of M the following inclusions

$$\text{Ass}(M) \subset \text{Supp}(\mathscr{F}) \subset \text{Supp}(M) \tag{1}$$

of sets hold.

Of course, $\text{Supp}(\mathscr{F}) \subset \text{Supp}(M)$. In order to show that $\text{Ass}(M) \subset \text{Supp}(\mathscr{F})$, let $P \in \text{Ass}(M)$. Then $PS_P \in \text{Ass}(M_P)$, so that $\text{depth}(M_P) = 0$. Hence, if $0 = M_0 \subset M_1 \subset \ldots \subset M_r = M$ with $M_i/M_{i-1} \cong (S/P_{F_i})(-a_i)$ is the prime filtration $\mathscr{F}$, then by applying the Depth Lemma [30, Proposition 1.2.9] we get

$$0 = \text{depth}(M_P) \geq \min\{\text{depth}((M_i/M_{i-1})_P): (M_i/M_{i-1})_P \neq 0\}.$$

Hence there exists an index i such that $\text{depth}(M_i/M_{i-1})_P = 0$. Therefore, $\text{depth}(S_P/P_i S_P) = 0$, and this is only possible if $P_i = P$.

3 Upper and Lower Bounds for Stanley Depth

The main purpose of this section is to prove the following inequalities

Theorem 9. *Let M be a finitely generated $\mathbb{Z}^n$-graded S-module, and let $\mathscr{F}$ be a prime filtration of M. Then*

$$\min\{\dim(S/P): P \in \text{Supp}(\mathscr{F})\} \leq \text{fdepth}(M) \leq \min\{\text{depth}(M), \text{sdepth}(M)\},$$

and

$$\max\{\text{depth}(M), \text{sdepth}(M)\} \leq \min\{\dim(S/P): P \in \text{Ass}(M)\} \leq \dim(M),$$

where for the validity of the upper bound for the Stanley depth we require that $\dim_K M_a \leq 1$ for all $a \in \mathbb{Z}^n$.

Proof. The lower bounds for the filtration depth follows from its definition. Let $\mathscr{F}_0 : 0 = M_0 \subset M_1 \subset \ldots \subset M_r = M$ be a prime filtration of M with

$\operatorname{supp}(\mathscr{F}_0) = \{P_1, \ldots, P_r\}$ and $\operatorname{fdepth}(M) = \min\{\dim(S/P_i): \ i = 1, \ldots, r\}$. By applying the depth lemma we obtain

$$\operatorname{fdepth}(M) = \min_i\{\dim(S/P_i)\} = \min_i\{\operatorname{depth}(M_i/M_{i-1})\} \leq \operatorname{depth}(M).$$

The inequality $\operatorname{fdepth}(M) \leq \operatorname{sdepth}(M)$ is trivial.

The inequality $\operatorname{depth}(M) \leq \min\{\dim(S/P): P \in \operatorname{Ass}(M)\}$ can be found in [30, Proposition 1.2.13], and the inequality $\min\{\dim(S/P): P \in \operatorname{Ass}(M)\} \leq \dim(M)$ is trivial.

Thus it remains to show that $\operatorname{sdepth}(M) \leq \min\{\dim(S/P): P \in \operatorname{Ass}(M)\}$. This inequality has first been shown by Apel in [13] in the case that $M = S/I$, where I is a monomial ideal. We follow his line of arguments to prove the general case. Let $\mathscr{D} : M = \bigoplus_{i=1}^{r} m_i K[Z_i]$ be a Stanley decomposition of M, and let $P \in \operatorname{Ass}(M)$. There exists a homogeneous element $m \in M$ with $P = \operatorname{Ann}(m)$. Since we assume that $\dim_K M_a \leq 1$ for all $a \in \mathbb{Z}^n$, it follows that $m \in m_i K[Z_i]$ for some i. Therefore, $x_j \notin P$ for all $x_j \in Z_i$. Hence

$$\operatorname{sdepth}(M) \leq |Z_i| \leq \dim(S/P).$$

This yields the desired inequality.

Question 10.
Is the inequality $\operatorname{sdepth}(M) \leq \min\{\dim(S/P): P \in \operatorname{Ass}(M)\} \leq \dim(M)$ true without the extra assumption that $\dim_K M_a \leq 1$ for all $a \in \mathbb{Z}^n$?

Generally speaking, Stanley decompositions of $\mathbb{Z}^n$-graded S-modules which admit graded components of dimension > 1 are not well understood. We will come back to this problem later.

Recall that M is said to be Cohen–Macaulay, if $\operatorname{depth}(M) = \dim(M)$. We call a module *pseudo Cohen–Macaulay*, if $\operatorname{sdepth}(M) = \dim(M)$. It follows from Theorem 9 that a pseudo Cohen–Macaulay module M with $\dim_K M_a \leq 1$ for all $a \in \mathbb{Z}^n$ has no embedded prime ideals, and assuming Stanley's conjecture holds, M is pseudo Cohen–Macaulay if it is Cohen–Macaulay.

Problem 11. Find a pseudo Cohen–Macaulay module which is not Cohen–Macaulay.

The inequality $\operatorname{fdepth}(M) \leq \operatorname{depth}(M)$ may be strict. To give such an example, we recall a few concepts regarding simplicial complexes and Stanley–Reisner rings. Let $[n] = \{1, \ldots, n\}$. A collection Δ of subsets of $[n]$ is called a *simplicial complex on the vertex set* $[n]$ if it satisfies the condition that whenever $F \in \Delta$ and $G \subset F$, then $G \in \Delta$. The elements of Δ are called *faces*. The set of facets (maximal faces under inclusion) of Δ is denoted $\mathscr{F}(\Delta)$.

We fix a field K. The *Stanley–Reisner ideal* $I_\Delta \subset S = K[x_1, \ldots, x_n]$ of Δ is the ideal generated be the squarefree monomials $\mathbf{x}_F$ with $F \notin \Delta$. Here $\mathbf{x}_F = x_{i_1} x_{i_2} \cdots x_{i_k}$ for $F = \{i_1, i_2, \ldots, i_k\}$. The factor ring $K[\Delta] = S/I_\Delta$ is called the

Stanley–Reisner ring of Δ. The dimension, $\dim(F)$, of a face F of Δ is the number $|F| - 1$, and the dimension of Δ is defined to be $\dim(\Delta) = \max\{\dim(F): F \in \Delta\}$. One has $\dim(K[\Delta]) = \dim(\Delta) + 1$. We refer the reader to [30] and [91] regarding basic properties of Stanley–Reisner ideals.

Now we come to our example. Let Δ be the simplicial complex on the vertex set $\{1, \ldots, 6\}$, associated to a triangulation of the real projective plane $\mathbb{P}^2$, whose facets are

$$\mathscr{F}(\Delta) = \{125, 126, 134, 136, 145, 234, 235, 246, 356, 456\}.$$

Here we write for simplicity $i_1 i_2 \ldots i_k$ for the set $\{i_1, i_2, \ldots, i_k\}$. The Stanley–Reisner ideal of Δ is

$$I_\Delta = (x_1 x_2 x_3, x_1 x_2 x_4, x_1 x_3 x_5, x_1 x_4 x_6, x_1 x_5 x_6,$$
$$x_2 x_3 x_6, x_2 x_4 x_5, x_2 x_5 x_6, x_3 x_4 x_5, x_3 x_4 x_6).$$

It is known that $\mathrm{depth}(I_\Delta) = 4$ if $\mathrm{char}\, K \neq 2$ and $\mathrm{depth}(I_\Delta) = 3$ if $\mathrm{char}\, K = 2$. Since the inequality $\mathrm{fdepth}(I_\Delta) \leq \mathrm{sdepth}(I_\Delta)$ holds independent of the characteristic of the base field, we obtain that $\mathrm{fdepth}(I_\Delta) \leq 3$. Therefore, $\mathrm{fdepth}(I_\Delta) < \mathrm{depth}(I_\Delta)$ for any field K with $\mathrm{char}\, K \neq 2$.

Problem 12. Find a $\mathbb{Z}^n$-graded S module such that, independent of the characteristic of K, $\mathrm{fdepth}(M) < \mathrm{depth}(M)$.

For some of the results described in the next section we recall a few more concepts related to simplicial complexes. For a subset of faces $F_1, \ldots, F_r$ of Δ we denote by $\langle F_1, \ldots, F_r \rangle$ the smallest subcomplex of Δ containing the faces $F_1, \ldots, F_r$. The simplicial complex Δ is called *shellable*, if the facets of Δ can be ordered $F_1, \ldots, F_r$ such that for $i = 2, \ldots, r$ the facets of $\langle F_1, \ldots, F_i \rangle \cap \langle F_i \rangle$ are maximal proper faces of $\langle F_i \rangle$. For $i \geq 2$ we denote by a_i the number of facets of $\langle F_1, \ldots, F_{i-1} \rangle \cap \langle F_i \rangle$, and set $a_1 = 0$. We call the $a_1, \ldots, a_r$ the sequence of *shelling numbers* of the given shelling of Δ.

Classically one requires from a shellable simplicial complex Δ that it is also pure. In that case $K[\Delta]$ is Cohen–Macaulay, see [30, Theorem 5.1.13] and [91, Theorem 8.2.6]. The definition of shellability given here which does not require that Δ is pure was first introduced by Björner and Wachs [22] and is sometimes called *nonpure shellability*. A shellable simplicial complex as defined here has the nice property that $K[\Delta]$ is sequentially Cohen–Macaulay, see [91, Corollary 8.2.19].

Finally we recall that the *Alexander dual* $\Delta^\vee$ of Δ is defined to be the simplicial complex whose faces are $\{[n] \setminus F: F \notin \Delta\}$. The Stanley–Reisner ideal of $\Delta^\vee$ is minimally generated by all monomials $x_{i_1} \cdots x_{i_k}$ where $(x_{i_1}, \ldots, x_{i_k})$ is a minimal prime ideal of I_Δ, see [91, Corollary 1.5.5].

4 Clean and Pretty Clean Modules

Let M be a finitely generated $\mathbb{Z}^n$-graded S-module. According to Dress [50] a prime filtration $\mathcal{F}$ of M is called *clean* if $\operatorname{Supp}(\mathcal{F}) = \operatorname{Min}(M)$, and M itself is called *clean* if M admits a clean filtration.

Since $\operatorname{Ass}(M) \subset \operatorname{Supp}(\mathcal{F})$, it follows that $\operatorname{Ass}(M) = \operatorname{Min}(M)$, if M is clean. In other words, a clean module has no embedded prime ideals. Moreover, if $\dim_K M_{\mathbf{a}} \leq 1$ for all $\mathbf{a} \in \mathbb{Z}^n$, then Theorem 9 implies that

$$\operatorname{fdepth}(M) = \operatorname{sdepth}(M) = \operatorname{depth}(M), \quad \text{if } M \text{ is clean.} \tag{2}$$

Hence clean modules satisfy Stanley's conjecture.

The question is whether clean modules appear in nature. At least for modules of the form S/I, where I is a squarefree monomial ideal, there is nice interpretation of cleanness. Since I is a squarefree monomial ideal, there exists a (unique) simplicial complex Δ on the vertex set $[n]$ such that $I = I_\Delta$, where I_Δ is the Stanley Reisner ideal of Δ.

Theorem 13 (Dress). *The simplicial complex Δ is shellable if and only if $K[\Delta]$ is a clean ring.*

Here we only show that when Δ is shellable, then $K[\Delta]$ is clean. More precisely we present the following result taken from [95]:

Proposition 14. *Let Δ be a shellable simplicial complex with shelling $F_1, \ldots, F_r$ and shelling numbers $a_1, \ldots, a_r$.*
  Then the filtration $(0) = M_0 \subset M_1 \subset \cdots \subset M_{r-1} \subset M_r = K[\Delta]$ of $K[\Delta]$ with

$$M_i = (\bigcap_{j=1}^{r-i} P_{F_j})/I_\Delta \quad \text{and} \quad M_i/M_{i-1} \cong S/P_{F_{r-i+1}}(-a_{r-i+1})$$

is a clean filtration of $K[\Delta]$.

Proof. Recall that $P_{F_i} = (\{x_j\}_{j \notin F_i})$, and $I_\Delta = \bigcap_{i=1}^r P_{F_i}$. Therefore, if $F_1, \ldots, F_r$ is a shelling of Δ, then for $i = 2, \ldots, r$ we have

$$\bigcap_{j=1}^{i-1} P_{F_j} + P_{F_i} = P_{F_i} + (f_i).$$

Here $f_i = \prod_k x_k$, where the product is taken over those $k \in F_i$ such that $F_i \setminus \{k\}$ is a facet of $\langle F_1, \ldots, F_{i-1}\rangle \cap \langle F_i\rangle$. In particular it follows that $\deg f_i$ equals the ith shelling number a_i.

We obtain the following isomorphisms of graded S-modules

$$(\bigcap_{j=1}^{i-1} P_{F_j})/(\bigcap_{j=1}^{i} P_{F_j}) \cong (\bigcap_{j=1}^{i-1} P_{F_j} + P_{F_i})/P_{F_i} = (P_{F_i} + (f_i))/P_{F_i}$$

$$\cong (f_i)/(f_i)P_{F_i} \cong S/P_{F_i}(-a_i).$$

The isomorphism $(P_{F_i} + (f_i))/P_{F_i} \cong (f_i)/(f_i)P_{F_i}$ results from the fact that $(f_i) \cap P_{F_i} = (f_i)P_{F_i}$ since the set of variables dividing f_i and the set of variables generating P_{F_i} have no element in common.

The theorem of Dress combined with the discussion preceding it shows that

$$\text{sdepth}(K[\Delta]) = \text{depth}(K[\Delta])$$

for any shellable simplicial complex.

The following results demonstrates the use of the theorem of Dress.

Proposition 15. *Let* $I \subset S$ *a squarefree monomial ideal which is a complete intersection. Then* S/I *is clean, and hence* $\text{fdepth}(S/I) = \text{sdepth}(S/I) = \text{depth}(S/I)$.

Proof. Let Δ be the simplicial complex whose Stanley–Reisner ideal I_Δ is equal to I.

Let $G(I) = \{v_1, \ldots, v_m\}$. Then $\text{supp}(v_i) \cap \text{supp}(v_j) = \emptyset$ for all $i \neq j$, since I is a complete intersection. Hence by what we said at the end of Sect. 3 it follows that $I_{\Delta^\vee}$ is minimally generated by the monomials of the form $x_{i_1} \ldots x_{i_m}$ where $x_{i_j} \in \text{supp}(v_j)$ for $j = 1, \ldots, m$. Thus we see that $I_{\Delta^\vee}$ is the matroidal ideal of the transversal matroid attached to the sets $\text{supp}(v_1), \ldots, \text{supp}(v_m)$, see [41, Sect. 5]. In [100, Lemma 1.3] and [41, Sect. 5] it is shown that any polymatroidal ideal has linear quotients, and this implies that Δ is a shellable simplicial complex, see [91, Proposition 8.2.5]. Hence by the theorem of Dress, S/I_Δ is clean. $\square$

Problem 16. Let Δ be a simplicial complex. Show that $\text{fdepth}(K[\Delta]) = \dim(K[\Delta])$ if and only if Δ is shellable.

In order to obtain a similar result as that of Dress for modules of the form S/I where I is a monomial ideal, but not necessarily squarefree, the definition of cleanness has been modified as follows [95].

Definition 17. Let M be a finitely generated $\mathbb{Z}^n$-graded S-module. A prime filtration $\mathscr{F}: 0 = M_0 \subset M_1 \subset M_2 \subset \cdots \subset M_r = M$ of M with $\text{Ass}(M_i/M_{i-1}) \cong S/P_i(-\mathbf{a}_i)$ for $i = 1, \ldots, r$ is called *pretty clean*, if for all $i < j$ for which $P_i \subset P_j$ it follows that $P_i = P_j$. In other words, a proper inclusion $P_i \subset P_j$ is only possible if $i > j$.

The module M is called *pretty clean*, if it has a pretty clean filtration. A ring is called pretty clean if it is a pretty clean module, viewed as a module over itself.

It is clear that any clean module is also pretty clean. It has been shown in [95, Theorem 10.5] that S/I, I a monomial ideal, is pretty clean if and only if the associated multi-complex is shellable. Hence the concept of pretty clean modules is the natural extension of clean modules. As for clean modules we have

Proposition 18. *Let* M *be a pretty clean module. Then*

$$\text{fdepth}(M) = \text{sdepth}(M) = \text{depth}(M).$$

Proof. Let $\mathscr{F}$ be a pretty clean filtration of M. We show that $\mathrm{Ass}(M) = \mathrm{Supp}(\mathscr{F})$. Then Theorem 9 yields the desired result.

Indeed, let $\mathscr{F}: 0 = M_0 \subset M_1 \subset M_2 \subset \cdots \subset M_r = M$ be a pretty clean filtration of M with $\mathrm{Ass}(M_i/M_{i-1}) \cong S/P_i(-a_i)$ for $i = 1, \ldots, r$. Now let $P_j \in \mathrm{Supp}(\mathscr{F})$. We may assume that $P_i \neq P_j$ for $i < j$. By localization we obtain the filtration

$$0 = (M_0)_{P_j} \subset (M_1)_{P_j} \subset (M_2)_{P_j} \subset \cdots \subset (M_r)_{P_j} = (M)_{P_j}$$

of M_{P_j} with $(M_i)_{P_j}/(M_{i-1})_{P_j} = (M_i/M_{i-1})_{P_j} = S_{P_j}/P_i S_{P_j} = 0$ for $i < j$. The last equation follows since $\mathscr{F}$ be a pretty clean filtration. It follows that $(M_j)_{P_j} = S_{P_j}/P_j S_{P_j}$. Therefore, $P_j \in \mathrm{Ass}(M_j)$, and hence $P_j \in \mathrm{Ass}(M)$ since $M_j \subset M$. This shows that $\mathrm{Supp}(\mathscr{F}) \subset \mathrm{Ass}(M)$. Since the other inclusion always holds, the claim follows.

There is a large and interesting class of ideals for which S/I is pretty clean. Let I and J be monomial ideals. We denote by $I : J^\infty$ the monomial ideal $\bigcup_{k \geq 1} I : J^k$. This ideal is called the *J-saturation* of I. A monomial ideal $I \subset S$ is called of *Borel type*, if

$$I : (x_j)^\infty = I : (x_1, \ldots, x_j)^\infty$$

for all $j = 1, \ldots, n$. Some authors call these ideals also ideals of nested type [19]. An important class of ideals of Borel type are the so-called Borel fixed ideals which play an important role in the theory of graded ideals, because they are just the generic initial ideals of graded ideals in a polynomial ring. Among the Borel fixed ideals are the strongly stable ideals.

Monomial ideals of Borel type can be characterized as follows [96, Proposition 2.2]:

Proposition 19. *Let I be a monomial ideal. The following conditions are equivalent:*

(a) *I is of Borel type.*
(b) *Let $u \in I$ be a monomial and suppose that $x_i^q | u$ for some $q > 0$. Then for all $j < i$ there exists an integer t such that $x_j^t (u/x_i^q) \in I$.*
(c) *Let $u \in I$ be a monomial; then for all integers i, j with $1 \leq j < i \leq n$ there exists an integer t such that $x_j^t (u/x_i^{v_i(u)}) \in I$, where $v_i(u)$ is the largest number for which $x_i^{v_i(u)}$ divides u.*

Proof. (a) $\Rightarrow$ (b): Let $u \in I$ be a monomial such that $x_i^q | u$ for some $q > 0$, and let $j < i$. Then $u = x_i^q v$ with $v \in I : (x_i)^\infty$. Condition (a) implies that $I : (x_i)^\infty \subset I : (x_j)^\infty$. Therefore, there exists t such that $x_j^t(u/x_i^q) = x_j^t v \in I$.

The implication (b) $\Rightarrow$ (c) is trivial. For the converse, let $u \in I$ be a monomial such that $x_i^q | u$ for some $q > 0$, and let $j < i$. By (c) there exists t such that $x_j^t(u/x_i^{v_i(u)}) \in I$. Therefore, $x_j^t(u/x_i^q) = x_i^{v_i(u)-q} x_j^t(u/x_i^{v_i(u)}) \in I$.

(b) $\Rightarrow$ (a): We will show that $I : x_i^\infty \subset I : (x_j)^\infty$ for $j < i$. This will imply (a). Let $u \in I : (x_i)^\infty$ be a monomial. Then $x_i^q u \in I$ for some $q > 0$, and so (b) implies that $x_j^t u \in I$ for some t, that is, $u \in I : (x_j)^\infty$.

For a monomial u we set $m(u) = \max\{i : v_i(u) \neq 0\}$, and for a monomial ideal $I \neq 0$ we set $m(I) = \max\{m(u) : u \in G(I)\}$.

Let $I \neq 0$ be a monomial ideal. Recursively we define an ascending chain of monomial ideals $I = I_0 \subset I_1 \subset \cdots$ as follows: We let $I_0 = I$. Suppose I_ℓ is already defined. If $I_\ell = S$, then the chain ends. Otherwise, we let $n_\ell = m(I_\ell)$, and set $I_{\ell+1} = I_\ell : (x_{n_\ell})^\infty$. Notice that $n \geq n_0 > n_1 > \cdots \geq 1$, so that this chain has length at most n. We call this chain of ideals, the *sequential chain of I*.

Lemma 20. *Let $I \neq 0$ be a monomial ideal of Borel type. Then*

$$I_{\ell+1} = I_\ell : (x_1, \ldots, x_{n_\ell})^\infty \quad \text{for all} \quad \ell.$$

Proof. We will show that $I_\ell : (x_{n_\ell})^\infty = I_\ell : (x_1, \ldots, x_{n_\ell})^\infty$ for all ℓ. Let $u \in I_\ell : (x_{n_\ell})^\infty$ be a monomial. Then there exists an integer q_ℓ such that $x_{n_\ell}^{q_\ell} u \in I_\ell$. Since $I_\ell = I_{\ell-1} : (x_{n_{\ell-1}})^\infty$, there exists an integer $q_{\ell-1}$ such that $x_{n_\ell}^{q_\ell} x_{n_{\ell-1}}^{q_{\ell-1}} u \in I_{\ell-1}$. Proceeding in this way we find integers $q_0, \ldots, q_\ell$ such that $x_{n_\ell}^{q_\ell} \cdots x_{n_0}^{q_0} u \in I$. By Proposition 19(c) there exists for all $j < n_\ell$ an integer t_j such that $x_j^{t_j} (x_{n_{\ell-1}}^{q_{\ell-1}} \cdots x_{n_0}^{q_0}) u \in I$. This implies that $x_j^{t_j} u \in I_\ell$ for all $j < n_\ell$. In other words, $u \in I_\ell : (x_1, \ldots, x_{n_\ell})^\infty$. $\quad\square$

Proposition 21. *Let I be a monomial ideal of Borel type. Then $R = S/I$ is pretty clean.*

Proof. We may assume that $I \neq 0$. Let $I = I_0 \subset I_1 \subset I_2 \subset \cdots \subset I_r = S$ be the sequential chain of I. Fix an integer $\ell < r$. Let $n_j = m(I_j)$ for all j, then the elements of $G(I_j)$ belong to $K[x_1, \ldots, x_{n_\ell}]$ for all $j \geq \ell$. Let J_l be the ideal generated by $G(I_\ell)$ in $K[x_1, \ldots, x_{n_\ell}]$. Then the saturation $J_\ell^{sat} = J_\ell : (x_1, \ldots, x_{n_\ell})^\infty$ is generated by the elements of $G(I_{\ell+1})$. It follows that

$$I_{\ell+1}/I_\ell = (J_\ell^{sat}/J_\ell)[x_{n_\ell+1}, \ldots, x_n]. \tag{3}$$

From this description it follows that $I_{\ell+1}/I_\ell$ has a prime filtration with the property that all of its factors are isomorphic to S/P with $P = (x_1, \ldots, x_{n_\ell})$. Composing the filtration of all the factors $I_{\ell+1}/I_\ell$ to a filtration of S/I, we see that S/I has a pretty clean filtration. $\quad\square$

5 Janet Decompositions

In this section we describe an algorithm which goes back to Maurice Janet [115], a French mathematician. Janet's algorithm applied to monomial ideals gives an explicit Stanley decomposition of these ideals. We call the Stanley decomposition obtained in this way the *Janet decomposition of I*.

In order to describe his construction, let $I \subset S = K[x_1, \ldots, x_n]$ be a monomial ideal. For each integer $k \geq 0$ we define the monomial ideals $I_k \subset S' = K[x_1, \ldots, x_{n-1}]$ by the equation

$$I \cap x_n^k K[x_1, \ldots, x_{n-1}] = I_k x_n^k.$$

It is clear that

$$I_0 \subset I_1 \subset I_2 \subset \ldots.$$

Since S' is Noetherian, we have that $I_k = I_{k+1}$ for $k \gg 0$. We define the integers

$$a = \min\{k \colon I_k \neq 0\},$$

and

$$b = \min\{d \colon I_k = I_{k+1} \text{ for all } k \geq d\}.$$

Then

$$I = \bigoplus_{a \leq k < b} I_k x_n^k \oplus I_b x_n^b K[x_n]. \tag{4}$$

Now Janet's decomposition is defined by induction on n. If $n = 1$, then $I = (x_1)^k$ and $I = x_1^k K[x_1]$ is the Janet decomposition of I. Suppose the Janet decomposition for any ideal in the polynomial ring S' is known. Let $I \subset S$ be a monomial ideal and consider its decomposition (4). By induction hypothesis we have for each I_k a Janet decomposition $I_k = \bigoplus_{j=1}^{r_k} u_{kj} K[Z_{kj}]$. By using (4) we obtain the Stanley decomposition

$$I = \bigoplus_{a \leq k < b} \bigoplus_{j=1}^{r_k} u_{kj} x_n^k K[Z_{kj}] \oplus \bigoplus_{j=1}^{r_b} u_{bj} x_n^b K[Z_{bj}, x_n].$$

The Stanley decomposition so obtained is called the *Janet decomposition* of I.

Example 22. Let $I = (x_1, x_2)^2 \subset K[x_1, x_2]$. Then with the notation introduced, $I_0 = x_1^2 K[x_1]$, $I_1 = x_1 K[x_1]$ and $I_k = K[x_1]$ for $k \geq 2$. Hence by (4), $I = x_1^2 K[x_1] \oplus (x_1 K[x_1]) x_2 \oplus (K[x_1]) x_2^2 K[x_2]$, which gives the Janet decomposition

$$I = x_1^2 K[x_1] \oplus x_1 x_2 K[x_1] \oplus x_2^2 K[x_1, x_2].$$

Problem 23. For any integer $k \geq 1$ compute the Janet decomposition of $(x_1, \ldots, x_n)^k$.

The following result is an immediate consequence of the Janet construction.

Corollary 24. *Let I be a monomial ideal in the polynomial in n variables $n \geq 1$. Then* $\text{sdepth}(I) \geq 1$.

A Janet decomposition can also be constructed for S/I where I is a monomial ideal. For this purpose we identify S/I as $\mathbb{Z}^n$-graded K-vector space with I^c whose monomial K-basis $\mathcal{B}$ consists of all monomials $u \in S \setminus I$. The monomial basis $\mathcal{B}$ of I^c forms an *order ideal* in the poset of all monomials ordered by divisibility, that is, if $u \in \mathcal{B}$ and $v|u$, then $v \in \mathcal{B}$. On the other hand, if $\mathcal{B}'$ is any order ideal, then the vector space spanned the monomials $u \in S \setminus \mathcal{B}'$ is a monomial ideal in S.

We observe that for all k there exist (uniquely determined) ideals $I_k \subset S'$ such that

$$I^c \cap x_n^k K[x_1, \ldots, x_{n-1}] = I_k^c x_n^k.$$

Since $I_0^c \supset I_1^c \supset \cdots$, it follows that $I_0 \subset I_1 \subset \cdots$ is an ascending chain of ideals in S' which stabilizes, because S' is Noetherian. Thus we may define the integers

$$a = \min\{k : I_k^c \neq 0\},$$

and

$$b = \min\{d : I_k^c = I_{k+1}^c \text{ for all } k \geq d\},$$

and obtain

$$I^c = \bigoplus_{a \leq k < b} I_k^c x_n^k \oplus I_b^c x_n^b K[x_n].$$

Equivalently, we obtain a $\mathbb{Z}^n$-graded direct sum decomposition of S/I, as an S'-module

$$S/I = \bigoplus_{a \leq k < b} (S'/I_k) x_n^k \oplus (S'/I_b) x_n^b \otimes_K K[x_n]. \tag{5}$$

As before, this is the first step in the inductive definition of a Janet decomposition of S/I. We leave it to the reader to complete the construction of the Janet decomposition for S/I. As an example the reader may work out the following

Problem 25. Compute the Janet decomposition of

$$S/(x_1^n, x_1^{n-1} x_2, x_1^{n-2} x_3, \cdots, x_1 x_n).$$

6   Stanley Depth and Regular Sequences

In this section we present a theorem of Rauf [164, Proposition 1.10] which describes how the Stanley depth of a $\mathbb{Z}^n$-graded module behaves under reduction modulo a variable which is regular on M.

Theorem 26. *Let M be a finitely generated $\mathbb{Z}^n$-graded S-module, and let x_k be regular on M. If $M/x_k M = \bigoplus_{i=1}^r n_i K[Z_i]$ is a Stanley decomposition of $M/x_k M$, where $n_i = m_i + x_k M$ and m_i is homogeneous, then $M = \bigoplus_{i=1}^r m_i K[Z_i, x_k]$ is a Stanley decomposition of M. In particular,* $\mathrm{sdepth}(M) \geq \mathrm{sdepth}(M/x_k M) + 1$.

Proof. Let $N = \sum_{i=1}^r m_i K[Z_i, x_k]$. Then $N \subset M$ and $M = N + x_k M$, by the choice of N. By induction on d one shows that $M = N + x_k^d M$. Indeed, suppose this is known for $d-1$. Then $M = N + x_k^{d-1}(N + x_k M) = N + x_k^{d-1} N + x_k^d M = N + x_k^d M$. From the fact that $M = N + x_k^d M$ for all d, one easily deduces $M = N$.

Suppose the sum $\sum_{i=1}^r m_i K[Z_i, x_k]$ is not direct. Then for $i = 1, \ldots, r$ there exist monomials $q_i \in K[Z_i, x_k]$, not all zero, such that $\sum_{i=1}^r m_i q_i = 0$. There exists a largest integer s such that x_k^s divides all q_i. Then $x_k^s(\sum_{i=1}^r m_i q_i') = 0$, where $q_i' = q_i/x_k^s$. Since x_k is a non-zero divisor on M it follows that $\sum_{i=1}^r m_i q_i' = 0$. This implies that $\sum_{i=1}^r n_i q_i' = 0$, which, since not all $q_i' = 0$, contradicts the assumption that $\bigoplus_{i=1}^r n_i K[Z_i]$ is a Stanley decomposition of $M/x_k M$.

It remains to be shown that each $m_i K[Z_i, x_k]$ is a Stanley space. Suppose that $m_i f = 0$ for some non-zero polynomial $f \in K[Z_i, x_k]$. We may assume that x_k does not divide f, since x_k is a non-zero divisor on M. Then $f = f_0 + f_1 x_k$ with $f_0 \in K[Z_i]$ and $f_0 \neq 0$. It follows that $n_i f_0 = 0$, and this is a contradiction since $n_i K[Z_i]$ is a Stanley space.

As a first application we have the following result which was stated in [36]. We present its proof following [31, Proposition 2.13].

Theorem 27. *Let M be a $\mathbb{Z}^n$-graded S-module. Then*

(a) $\mathrm{depth}(M) = 0$ *if* $\mathrm{sdepth}(M) = 0$;
(b) $\mathrm{sdepth}(M) = 0$ *if* $\mathrm{depth}(M) = 0$ *and* $\dim_K M_{\mathbf{a}} \leq 1$ *for all* $\mathbf{a}$.

Proof. (a) We show that if $\mathrm{depth}(M) > 0$, then $\mathrm{sdepth}(M) > 0$. We set $U_{n+1} = M$, $U_0 = 0$, and $U_i = \{m \in U_{i+1} : x_i^k m = 0 \text{ for some } k > 0\}$ for $i = 1, \ldots, n$. Then we obtain the following chain of $\mathbb{Z}^n$-graded submodules

$$0 = U_0 \subset U_1 \subset U_2 \subset \cdots \subset U_{n+1} = M.$$

This implies that $\mathrm{sdepth}(M) \geq \min\{\mathrm{sdepth}(U_{i+1}/U_i) : i = 0, \ldots, n\}$.

Since $U_1/U_0 = U_1 = \{m \in M : (x_1, \ldots, x_n)^k m = 0 \text{ for some } k\}$, and since $\mathrm{depth}(M) > 0$ it follows that $U_1/U_0 = 0$. By the definition of U_{i+1}/U_i we have that x_i is regular on U_{i+1}/U_i for $i > 0$. Thus Theorem 26 implies $\mathrm{sdepth}(U_{i+1}/U_i) > 0$ for $i > 0$. This yields the desired conclusion.

(b) Assuming that $\mathrm{depth}(M) = 0$, there exists a homogeneous element $m \neq 0$ such that $x_j m = 0$ for all j. Let $\mathscr{D}: M = \bigoplus_{i=1}^r m_i K[Z_i]$ be a Stanley decomposition with $\mathrm{sdepth}(\mathscr{D}) = \mathrm{sdepth}(M)$. Since $\dim_K M_{\mathbf{a}} \leq 1$ for all $\mathbf{a}$, it follows that $m \in m_i K[Z_i]$ for some i. However since $m_i K[Z_i]$ is a free $K[Z_i]$-submodule of M and since $x_j m = 0$ for all j, this is only possible if $K[Z_i] = K$. Thus $\mathrm{sdepth}(M) = 0$.

As a second application we obtain the following result of [62, Theorem 2.2], see also [31, Corollary 2.12].

Theorem 28. *Let M be a $\mathbb{Z}^n$-graded S-module, and*

$$\mathbb{F}: \cdots \to F_p \to F_{p-1} \to \cdots \to F_1 \to F_0 \to M \to 0$$

a $\mathbb{Z}^n$-graded free resolution of M. Let Z_p be the pth syzygy module of M with respect to this resolution, i.e., the image of $F_p \to F_{p-1}$. Then $\operatorname{sdepth}(Z_p) \geq \min\{p, n\}$.

Proof. One easily proves by induction on p that $x_1, \ldots, x_p$ is a regular sequence on Z_p. Thus the assertion follows immediately from Theorem 26.

7 How to Compute the Stanley Depth

In this section we present an algorithm, introduced in [101], in order to compute the Stanley depth of a module of the form I/J where $J \subset I \subset S$ are monomial ideals. The algorithm amounts to attach to each partition of a certain posets associated with I/J a Stanley decompositions of I/J, and to show that among the finitely many Stanley decompositions so obtained there is one which gives the Stanley depth of I/J. Here in this presentation, the proofs given are more straightforward than those in [102] where Miller's functors as defined in [140] are used.

We define a natural partial order on $\mathbb{N}^n$ as follows: $a \leq b$ if and only if $a(i) \leq b(i)$ for $i = 1, \ldots, n$. Note that $x^a | x^b$ if and only if $a \leq b$. Here, for any $c \in \mathbb{N}^n$ we denote as usual by x^c the monomial $x_1^{c(1)} x_2^{c(2)} \cdots x_n^{c(n)}$. Observe that $\mathbb{N}^n$ with the partial order introduced is a distributive lattice with meet $a \wedge b$ and join $a \vee b$ defined as follows: $(a \wedge b)(i) = \min\{a(i), b(i)\}$ and $(a \vee b)(i) = \max\{a(i), b(i)\}$. We also denote by ϵ_j the jth canonical unit vector in $\mathbb{Z}^n$.

Suppose I is generated by the monomials $x^{a_1}, \ldots, x^{a_r}$ and J by the monomials $x^{b_1}, \ldots, x^{b_s}$. We choose $g \in \mathbb{N}^n$ such that $a_i \leq g$ and $b_j \leq g$ for all i and j, and let $P_{I/J}^g$ be the set of all $c \in \mathbb{N}^n$ with $c \leq g$ and such that $a_i \leq c$ for some i and $c \not\geq b_j$ for all j. The set $P_{I/J}^g$ viewed as a subposet of $\mathbb{N}^n$ is a finite poset. We call it the *characteristic poset* of I/J with respect to g. There is a natural choice for g, namely the join of all the a_i and b_j. For this g, the poset $P_{I/J}^g$ has the least number of elements, and we denote it simply by $P_{I/J}$. Note that if Δ is a simplicial complex on the vertex set $[n]$, then P_{S/I_Δ} is just the face poset of Δ.

Figure 2 shows the characteristic poset for the maximal ideal $\mathfrak{m} = (x_1, x_2, x_3) \subset K[x_1, x_2, x_3]$. The elements of this poset correspond to the squarefree monomials x_1, x_2, x_3, $x_1 x_2$, $x_1 x_3$, $x_2 x_3$ and $x_1 x_2 x_3$. Thus the corresponding labels in Fig. 2 should be $(1, 0, 0), (0, 1, 0), \ldots, (1, 1, 1)$. In the squarefree case, like in this example, it is however more convenient and shorter to replace the $(0, 1)$-vectors (which label the vertices in the characteristic poset) by their support. In other words, each $(0, 1)$-vector with support $\{i_1 < i_2 < \cdots < i_k\}$ is replaced by $i_1 i_2 \cdots i_k$, as done in Fig. 2.

Given any poset P and $a, b \in P$ we set $[a, b] = \{c \in P: a \leq c \leq b\}$ and call $[a, b]$ an *interval*. Of course, $[a, b] \neq \emptyset$ if and only if $a \leq b$. Suppose P is a finite poset. A *partition* of P is a disjoint union

Fig. 2 A poset P

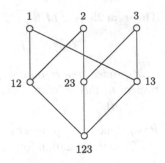

Fig. 3 A partition of the poset P

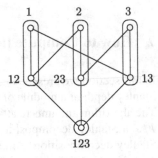

$$\mathscr{P}: \; P = \bigcup_{i=1}^{r} [a_i, b_i]$$

of intervals.

Figure 3 displays a partition of the poset given in Fig. 2. The framed regions in Fig. 3 indicate that $P_{\mathrm{m}} = [1, 12] \cup [2, 23] \cup [3, 13] \cup [123, 123]$.

We will show that each partition of $P_{I/J}^g$ gives rise to a Stanley decomposition of I/J.

In order to describe the Stanley decomposition of I/J coming from a partition of $P_{I/J}^g$ we shall need the following notation: for each $b \in P_{I/J}^g$, we set $Z_b = \{x_j: b(j) = g(j)\}$. We also introduce the function

$$\rho: P_{I/J}^g \to \mathbb{Z}_{\geq 0}, \quad c \mapsto \rho(c),$$

where $\rho(c) = |\{j: c(j) = g(j)\}|(= |Z_c|)$. We then have

Theorem 29. (a) *Let* $\mathscr{P}: P_{I/J}^g = \bigcup_{i=1}^{r} [c_i, d_i]$ *be a partition of* $P_{I/J}^g$. *Then*

$$\mathscr{D}(\mathscr{P}): I/J = \bigoplus_{i=1}^{r} (\bigoplus_{c} x^c K[Z_{d_i}]) \tag{6}$$

is a Stanley decomposition of I/J, *where the inner direct sum is taken over all* $c \in [c_i, d_i]$ *for which* $c(j) = c_i(j)$ *for all* j *with* $x_j \in Z_{d_i}$. *Moreover,*

$$\mathrm{sdepth}(\mathscr{D}(\mathscr{P})) = \min\{\rho(d_i): i = 1, \ldots, r\}.$$

(b) *One has*

$$\text{sdepth}(I/J) = \max\{\text{sdepth}(\mathcal{D}(\mathcal{P})): \ \mathcal{P} \text{ is a partition of } P^g_{I/J}\}.$$

In particular, there exists a partition $\mathcal{P}$: $P^g_{I/J} = \bigcup_{i=1}^r [c_i, d_i]$ *of* $P^g_{I/J}$ *such that*

$$\text{sdepth}(I/J) = \min\{\rho(d_i): \ i = 1, \ldots, r\}.$$

Proof. (a) We first show that the sum of the K-vector spaces in (6) is equal to the K-vector space spanned by all monomials $u \in I \setminus J$ (which of course is isomorphic to the K-vector space I/J).

Let $u = x^e$ be a monomial in $I \setminus J$ and let $c' = e \wedge g$. Then, $c' \in P^g_{I/J}$ and consequently, there exists $i \in \{1, \ldots, r\}$ such that $c' \in [c_i, d_i]$. Let c be the vector with

$$c(j) = \begin{cases} c_i(j), & \text{if } d_i(j) = g(j), \\ c'(j), & \text{otherwise.} \end{cases}$$

It follows from the definition of c that $x^c K[Z_{d_i}]$ is one of the Stanley spaces appearing in (6). We claim that $u \in x^c K[Z_{d_i}]$, equivalently, that $x^{e-c} \in K[Z_{d_i}]$. Indeed, if $x_j \in Z_{d_i}$, then $d_i(j) = g(j)$, and hence $e(j) \geq c'(j) \geq c_i(j) = c(j)$. On the other hand, if $x_j \notin Z_{d_i}$, then $g(j) > d_i(j) \geq c'(j) = c(j)$. Since $c'(j) = \min\{e(j), g(j)\}$, it therefore follows that $e(j) = c(j)$, as desired.

In order to prove that the sum (6) is direct, it suffices to show that any two different Stanley spaces in (6) have no monomial in common. Suppose to the contrary that $x^b \in x^p K[Z_{d_i}] \cap x^q K[Z_{d_j}]$ and that $x^p K[Z_{d_i}] \neq x^q K[Z_{d_j}]$ are both summands in (6). Since each of the inner sums in (6) is direct, we have that $i \neq j$.

We claim that $x^b \in x^p K[Z_{d_i}]$ yields $b \wedge g \in [c_i, d_i]$. Indeed, since $c_i \leq b \wedge g$, the claim follows once it is shown that $b \wedge g \leq d_i$. If $d_i(j) = g(j)$, then

$$(b \wedge g)(j) = \min\{b(j), g(j)\} \leq g(j) = d_i(j).$$

If $d_i(j) < g(j)$, then $x_j \notin Z_{d_i}$ and hence $b(j) = p(j)$. Together with the inequality $p(j) \leq d_i(j) < g(j)$, we obtain that $(b \wedge g)(j) = p(j) \leq d_i(j)$. In both cases the claim follows.

Similarly, since $x^b \in x^q K[Z_{d_j}]$ we see that $b \wedge g \in [c_j, d_j]$. This is a contradiction, since $[c_i, d_i] \cap [c_j, d_j] = \emptyset$.

The statement about the Stanley depth of $\mathcal{D}(\mathcal{P})$ follows immediately from the definitions.

(b) Let $\mathcal{D}$ be an arbitrary Stanley decomposition of I/J. We are going to construct a partition $\mathcal{P}$ of $P^g_{I/J}$ such that $\text{sdepth}(\mathcal{D}(\mathcal{P})) \geq \text{sdepth}(\mathcal{D})$. This will then yield the desired conclusion. First, to each $b \in P^g_{I/J}$ we assign an interval

$[c, d] \subset P_{I/J}^g$: since $x^b \in I \setminus J$, there exists a Stanley space $x^c K[Z]$ in the decomposition $\mathcal{D}$ of I/J with $x^b \in x^c K[Z]$. It follows that $c \in P_{I/J}^g$ and $b(j) = c(j)$ for all j with $x_j \notin Z$. Now, we define $d \in \mathbb{N}^n$ by setting

$$d(j) = \begin{cases} g(j), & \text{if } x_j \in Z, \\ c(j), & \text{if } x_j \notin Z. \end{cases}$$

Observe that $[c, d] \subset P_{I/J}^g$. We noticed already that $c \in P_{I/J}^g$. It remains to be shown that $d \in P_{I/J}^g$. Since $x^c K[Z] \in I \setminus J$, it follows that $x^{c + \sum_j n_j \epsilon_j} \in I \setminus J$, where the sum is taken over all j with $x_j \in Z$ and where for all j we have $n_j \in \mathbb{Z}_{\geq 0}$. Therefore $d = c + \sum_j (g(j) - c(j)) \epsilon_j \in P_{I/J}^g$.

Next we show that $b \in [c, d]$. For this we need to show that $b \leq d$. Indeed, if $x_j \in Z$, then $b(j) \leq g(j) = d(j)$. Otherwise $d(j) = c(j) = b(j)$ and consequently the inequality holds. Since $b \in [c, d]$, we obtain that $x^b \in x^c K[Z_d]$, and $Z \subseteq Z_d$, according to the definition of d.

In order to complete the proof we now show that the intervals constructed above provide a partition $\mathcal{P}$ of $P_{I/J}^g$ and that $\mathrm{sdepth}(\mathcal{D}(\mathcal{P})) \geq \mathrm{sdepth}(\mathcal{D})$.

It is clear that these intervals cover $P_{I/J}^g$. Therefore it is enough to check that for any $b_1, b_2 \in P_{I/J}^g$ with $b_1 \neq b_2$, the corresponding intervals obtained from our construction, say $[c_1, d_1]$ and $[c_2, d_2]$, satisfy either $[c_1, d_1] = [c_2, d_2]$ or $[c_1, d_1] \cap [c_2, d_2] = \emptyset$.

To each c_i corresponds a Stanley space $x^{c_i} K[Z_i]$ in the given Stanley decomposition $\mathcal{D}$. We consider two cases. In the first case, we assume that $c_1 = c_2$. Then $Z_1 = Z_2$, and consequently $d_1 = d_2$. Hence $[c_1, d_1] = [c_2, d_2]$. In the second case, we assume $c_1 \neq c_2$. In this case we prove that $[c_1, d_1] \cap [c_2, d_2] = \emptyset$. Assume, by contradiction, that there exists $e \in P_{I/J}^g$ such that $e \in [c_1, d_1] \cap [c_2, d_2]$. It follows from the construction of the interval $[c_1, d_1]$ that $c_1(j) = d_1(j)$ if $x_j \notin Z_1$. Therefore, $e \in [c_1, d_1]$ implies that $e(j) = c_1(j)$, for all j with $x_j \notin Z_1$, and hence we obtain that $x^e \in x^{c_1} K[Z_1]$. Analogously, one obtains that $x^e \in x^{c_2} K[Z_2]$, a contradiction since $x^{c_1} K[Z_1] \cap x^{c_2} K[Z_2] = 0$.

To establish the inequality $\mathrm{sdepth}(\mathcal{D}(\mathcal{P})) \geq \mathrm{sdepth}(\mathcal{D})$, we observe that $\mathrm{sdepth}(\mathcal{D}(\mathcal{P}))$ is equal to the minimum of all integers $|Z_d|$ where $[c, d]$ belongs to $\mathcal{P}$. On the other hand, we have already shown that for each Stanley space $x^c K[Z]$ in $\mathcal{D}$ such that $c \in P_{I/J}^g$ we have $|Z_d| \geq |Z|$. This yields the desired inequality.

Problem 30. Let $J \subset I$ be monomial ideals of $S = K[x_1, \ldots, x_n]$, and let $T = S[x_{n+1}]$ be the polynomial ring over S in the variable x_{n+1}. Then $\mathrm{sdepth}(IT/JT) = \mathrm{sdepth}(I/J) + 1$.

Problem 31. Use Theorem 29 to show that if I is a squarefree monomial ideal, then $\mathrm{sdepth}(I) \geq \min\{\deg(u): u \in G(I)\}$.

As a first application we will show that Stanley's conjecture on the Stanley depth implies another conjecture by Stanley regarding partitions of simplicial complexes.

Let Δ be a simplicial complex of dimension $d - 1$ on the vertex set $[n]$. A subset $\mathscr{I} \subset \Delta$ is called an *interval*, if there exist faces $F, G \in \Delta$ such that $\mathscr{I} = \{H \in \Delta : F \subseteq H \subseteq G\}$. We denote this interval given by F and G also by $[F, G]$. A *partition* $\mathscr{P}$ of Δ is a presentation of Δ as a disjoint union of intervals.

Let $I_\Delta \subset S$ be the Stanley–Reisner ideal of Δ and $K[\Delta] = S/I_\Delta$ its Stanley–Reisner ring. Obviously partitions of $P_{K[\Delta]}^g$ with $g = (1, 1, \ldots, 1)$ correspond to partitions of Δ. To simplify notation we write $P_{K[\Delta]}$ for $P_{K[\Delta]}^g$ when $g = (1, \ldots, 1)$, and similarly P_{I_Δ} for $P_{I_\Delta}^g$.

A Stanley space $uK[Z]$ is called a *squarefree Stanley space*, if u is a squarefree monomial and $\mathrm{supp}(u) \subseteq Z$. We shall use the following notation: for $F \subseteq [n]$ we set $x_F = \prod_{i \in F} x_i$ and $Z_F = \{x_i : i \in F\}$. Then a Stanley space is squarefree if and only if it is of the form $x_F K[Z_G]$ with $F \subseteq G \subseteq [n]$.

A Stanley decomposition of S/I is called a *squarefree Stanley decomposition* of S/I if all Stanley spaces in the decomposition are squarefree.

Having in mind the relationship between partitions of simplicial complexes and partitions of the characteristic poset of the corresponding Stanley–Reisner ideal, we obtain as an immediate consequence of Theorem 29 the following

Corollary 32. *Let Δ be a simplicial complex on the vertex set $[n]$, and let $\mathscr{P} \colon \Delta = \bigcup_{i=1}^{r} [F_i, G_i]$ be a partition of Δ. Then*

(a) $\mathscr{D}(\mathscr{P}) \colon K[\Delta] = \bigoplus_{i=1}^{r} x_{F_i} K[Z_{G_i}]$ *is a squarefree Stanley decomposition of $K[\Delta]$.*

(b) *The map $\mathscr{P} \mapsto \mathscr{D}(\mathscr{P})$ establishes a bijection between partitions of Δ and squarefree Stanley decompositions of $K[\Delta]$.*

(c) *There exists a partition $\mathscr{P}$ of Δ such that* $\mathrm{sdepth}(K[\Delta]) = \mathrm{sdepth}(\mathscr{D}(\mathscr{P}))$.

It follows from statement (c) of the preceding corollary that Stanley's conjecture holds for the Stanley–Reisner ring $K[\Delta]$ if and only if there exists a partition $\Delta = \bigcup_{i=1}^{r} [F_i, G_i]$ of Δ with $|G_i| \geq \mathrm{depth}(K[\Delta])$ for all i.

Let $\mathscr{F}(\Delta)$ denote the set of facets of Δ. Stanley calls a simplicial complex Δ *partitionable* if there exists a partition $\Delta = \bigcup_{i=1}^{r} [F_i, G_i]$ with $\mathscr{F}(\Delta) = \{G_1, \ldots, G_r\}$. Stanley conjectures [182, Conjecture 2.7] (see also [183, Problem 6]) that each Cohen–Macaulay simplicial complex is partitionable. In view of Corollary 32(c) and the comments following it, we see that the conjecture on Stanley depth implies the conjecture on partitionable simplicial complexes.

In the next section we will see that Stanley's conjecture on the Stanley depth, restricted to Stanley–Reisner rings, is indeed equivalent to his conjecture on partionable simplicial complexes.

8 Characteristic Posets and Skeletons

The main purpose of this section is the proof of the fact, shown in [99], that Stanley's conjecture holds for all S/I, I a monomial ideal, if and only if it holds for all such algebras S/I which are Cohen–Macaulay.

Let $J \subset I \subset S$ be monomial ideals. We choose $g \in \mathbb{N}^n$ for which the characteristic poset $P_{I/J}^g$ of I/J is defined. In the previous section we introduced the ρ-function, whose definition we now extend as follows: for any $b \in \mathbb{N}^n$ we define subsets $Y_b = \{x_j : b(j) \neq g(j)\}$ and $Z_b = \{x_j : b(j) = g(j)\}$ of $X = \{x_1, \ldots, x_n\}$ and set $\rho(b) = |\{j : b(j) = g(j)\}| = |Z_b|$.

Lemma 33. *With the notation introduced we have*

(a) $\dim(I/J) = \max\{\rho(b) : b \in P_{I/J}^g\}$.

(b) $\rho(b) \leq \dim(I/J)$ *for all* $b \in \mathbb{N}^n$ *with* $x^b \in I \setminus J$.

Proof. (a) We choose a partition $\mathscr{P} : P_{I/J}^g = \bigcup_{i=1}^r [c_i, d_i]$ of $P_{I/J}^g$. Then Theorem 29 yields the Stanley decomposition

$$\mathscr{D}(\mathscr{P}) : I/J = \bigoplus_{i=1}^r (\bigoplus_c x^c K[Z_{d_i}]) \tag{7}$$

of I/J, where the inner direct sum is taken over all $c \in [c_i, d_i]$ for which $c(j) = c_i(j)$ for all j with $x_j \in Z_{d_i}$.

We use this Stanley decomposition to compute the Hilbert series of I/J and obtain

$$\mathrm{Hilb}(I/J) = \sum_{i=1}^r \sum_c \frac{t^{|c|}}{(1-t)^{\rho(d_i)}}$$

with the summation on the c as above. Here $|c| = \sum_{i=1}^n c_i$.
We may assume that $\rho(d_r) \geq \rho(d_i)$ for all i. Then

$$\mathrm{Hilb}(I/J) = \frac{\sum_{i=1}^r \sum_c t^{|c|} (1-t)^{\rho(d_r) - \rho(d_i)}}{(1-t)^{\rho(d_r)}} = \frac{Q(t)}{(1-t)^{\rho(d_r)}}$$

with $Q(1) \neq 0$. Thus it follows from [91, Theorem 6.1.3] that $\rho(d_r) = \dim(I/J)$. Since $\rho(d_r) = \max\{\rho(b) : b \in P_{J/I}^g\}$, the assertion follows.

(b) Let $g' = g \vee b$. Then $b \in P_{I/J}^{g'}$, and hence $|\{j : b(j) = g'(j)\}| \leq \dim(I/J)$, by (a). Since $\rho(b) = |\{j : b(j) = g(j)\}| \leq |\{j : b(j) \geq g(j)\}| = |\{j : b(j) = g'(j)\}|$, the assertion follows.

Motivated by Lemma 33(a) we consider for each $j \leq d$, the monomial ideal I_j generated by I together with all monomials x^b such that $\rho(b) > j$. We then obtain a chain of monomial ideals

$$I = I_d \subset I_{d-1} \subset \cdots \subset I_0 \subset S.$$

Of course this chain of ideals depends not only on I, but also on the choice of g.

Consider the special case, where $I = I_\Delta$ is the Stanley–Reisner ideal of a simplicial complex Δ on the vertex set $\{1, \ldots, n\}$. Then for $g = (1, \ldots, 1)$ we have $I_j = I_{\Delta^{(j)}}$ where $\Delta^{(j)}$ is the jth skeleton of Δ, that is the subcomplex of Δ with faces $F \in \Delta$ with $\dim(F) \leq j$. This observation justifies to call I_j the jth *skeleton ideal* of I (with respect to g).

Theorem 34. *For each* $0 \leq j \leq d$, *the factor module* I_{j-1}/I_j *is a direct sum of cyclic Cohen–Macaulay modules of dimension* j. *In particular,* I_{j-1}/I_j *is a* j-*dimensional Cohen–Macaulay module.*

Proof. Replacing I by I_j it suffices to consider the case $j = d$. Let

$$J = (I, \{x^b : b \in A\}), \quad \text{where} \quad A = \{b \in P^g_{S/I} : \rho(b) = d\},$$

then $I_{d-1}/I_d = J/I$.

Let $\{Z_1, \ldots, Z_r\}$ be the collection of those subsets of X with the property that for each $i = 1, \ldots, r$ there exists $b \in A$ such that $Z_i = Z_b$. Let $A_i = \{b \in A : Z_b = Z_i\}$, and let $b, b' \in A_i$. Then $b \wedge b' \in A_i$. Thus the meet of all the elements in A_i is the unique smallest element in A_i. We denote this element by b_i. Then $Z_i = Z_{b_i}$. Obviously the elements $f_i = x^{b_i} + I, i = 1, \ldots, r$ generate I_{d-1}/I_d. We claim that

$$I_{d-1}/I_d = \bigoplus_{i=1}^r S f_i.$$

The cyclic module $S f_i$ is $\mathbb{Z}^n$-graded with a K-basis $x^a + I$ with $a \geq b_i$ and $x^a \notin I$. Given $c \in \mathbb{N}^n$ with $c \geq b_i$ and $c \geq b_j$ for some $1 \leq i < j \leq r$, then $\rho(c) > d$, and so $x^c \in I$, by Lemma 33. This shows that the sum of the cyclic modules $S f_i$ is indeed direct.

Next we notice that if $x^c = x^{c_1} x^{c_2}$ with $x^{c_1} \in K[Z_{b_i}]$ and $x^{c_2} \in K[Y_{b_i}]$ belong to $\mathrm{Ann}(S f_i)$, then $x^{c_2} \in \mathrm{Ann}(S f_i)$. Indeed, $x^c = x^{c_1} x^{c_2} \in \mathrm{Ann}(S f_i)$ if and only if $a_j \leq b_i + c_1 + c_2$ for some j. Since $c_1(k) = 0$ for all k with $x_k \in Y_{b_i}$, it follows that $a_j(k) \leq (b_i + c_2)(k)$ for all $k \in Y_{b_i}$, while for k with $x_k \in Z_{b_i}$ we have $a_j(k) \leq g(k) = b_i(k) = (b_i + c_2)(k)$. Hence $a_j \leq b_i + c_2$, which implies that $x^{c_2} \in \mathrm{Ann}(S f_i)$.

It follows that $\mathrm{Ann}(S f_i)$ is generated by monomials in $K[Y_{b_i}]$. In other words, there exists a monomial ideal $M_i \subset K[Y_{b_i}]$ such that $\mathrm{Ann}(S f_i) = M_i S$.

For each k with $x_k \in Y_{b_i}$ we have $b_i(k) < g(k)$ and $\rho(b_i + (g(k) - b_i(k))\epsilon_k) = d + 1$. Therefore Lemma 33 implies that $x^{b_i} x_k^{g(k)-b_i(k)} \in I$. It follows that $x_k^{g(k)-b_i(k)} \in M_i$ for all k with $x_k \in Y_{b_i}$. Hence we see that $\dim(K[Y_{b_i}]/M_i) = 0$. This implies that $S f_i = S/M_i S$ is Cohen–Macaulay of dimension d.

As a first application of Theorem 34 we obtain the following characterization of the depth of S/I which generalizes a classical result of Hibi [104, Corollary 2.6].

Corollary 35. *Let $I \subset S$ be a monomial ideal. Then*

$$\operatorname{depth}(S/I) = \max\{j : S/I_j \text{ is Cohen–Macaulay}\},$$

and S/I_j is Cohen–Macaulay for all $j \le \operatorname{depth}(S/I)$.

Proof. Let $d = \dim(S/I)$ and $t = \operatorname{depth}(S/I)$. Since $I_j = (I_{d-1})_j$ for $j \le d-1$, both assertions follow by induction on d once we can show the following:

(i) If $t < d$, then $\operatorname{depth}(S/I_{d-1}) = t$.
(ii) If S/I is Cohen–Macaulay, then S/I_{d-1} is Cohen–Macaulay.

Proof of (i): The exact sequence

$$0 \longrightarrow I_{d-1}/I \longrightarrow S/I \longrightarrow S/I_{d-1} \longrightarrow 0$$

implies that

$$\operatorname{depth}(S/I_{d-1}) \ge \min\{\operatorname{depth}(I_{d-1}/I) - 1, \operatorname{depth}(S/I)\}, \tag{8}$$

with equality if $t < d - 1$, see [30, Proposition 1.2.9]. By Theorem 34, $\operatorname{depth}(I_{d-1}/I) - 1 = d - 1$. It follows that $\operatorname{depth}(S/I_{d-1}) = t$, if $t < d - 1$. On the other hand, if $t = d - 1$, then $\operatorname{depth}(S/I_{d-1}) \ge d - 1$. However, since $\dim(S/I_{d-1}) = d - 1$, we again get $\operatorname{depth}(S/I_{d-1}) = d - 1 = t$.

Proof of (ii): If S/I is Cohen–Macaulay, then $\operatorname{depth}(S/I) = d$. Hence Theorem 34 and inequality (8) imply that $\operatorname{depth}(S/I_{d-1}) \ge d - 1$. Since $\dim(S/I_{d-1}) = d - 1$, the assertion follows.

Now we come to the application to Stanley depth.

Proposition 36. *For all $0 \le j \le d = \dim(S/I)$ we have*

$$\operatorname{sdepth}(S/I) \ge \operatorname{sdepth}(S/I_j).$$

Proof. Observe that $P^g_{S/I_j} = \{a \in P^g_{S/I} : \rho(a) \le j\}$. Let t be the Stanley depth of S/I_j. By Theorem 29 there exists a partition $\mathscr{P} \colon P^g_{S/I_j} = \bigcup_{i=1}^r [c_i, d_i]$ with $\rho(\mathscr{P}) = t$. We complete the partition of $\mathscr{P}$ to a partition $\mathscr{P}'$ of S/I by adding the intervals $[a, a]$ with $a \in P^g_{S/I} \setminus P^g_{S/I_j}$. Since $\rho(a) > j$ for all $a \in P^g_{S/I} \setminus P^g_{S/I_j}$ it follows that $\rho(\mathscr{P}') = t$. Hence, again by Theorem 29, $\operatorname{sdepth}(S/I) \ge t$, as desired.

We denote by $\mathscr{A}_n$ the set of all K-algebras of the form S/I where $S = K[x_1, \ldots, x_n]$ and I is a monomial ideal.

Corollary 37. *Suppose Stanley's conjecture holds for all Cohen–Macaulay K-algebras in $\mathscr{A}_n$. Then Stanley's conjecture holds for all K-algebras in $\mathscr{A}_n$.*

Proof. Let I be a monomial ideal and suppose that $t = \operatorname{depth}(S/I)$. Then S/I_t is Cohen–Macaulay of dimension t, see Corollary 35. Our assumption implies that $\operatorname{sdepth}(S/I_t) = t$. Thus the assertion follows from Proposition 36.

Remark 38. Combining the above Corollary 37 with the equations of (2) and with Theorem 13 we see that if there exists a simplicial complex Δ for which $K[\Delta]$ Stanley's conjecture does not hold, then there exists such a such simplicial complex Δ with the additional property that $K[\Delta]$ is Cohen–Macaulay, and this simplicial complex Δ cannot be shellable. Thus possible counterexamples to Stanley's conjecture should be searched among the non-shellable Cohen–Macaulay simplicial complexes

9   Special Partitions of Posets

In [102] it was conjectured that $\mathrm{sdepth}(\mathfrak{m}) = \lceil n/2 \rceil$ where $\mathfrak{m} = (x_1, \ldots, x_n)$ is the graded maximal ideal of $S = K[x_1, \ldots, x_n]$ and where $\lceil n/2 \rceil$ denotes smallest integer $\geq n/2$. This conjecture has been proved by Biró et al. [18]. We will present their beautiful solution here and refer to their paper for the details of the proof.

According to Theorem 29 one has to find a partition of $\mathscr{P}: P_\mathfrak{m} = \bigcup_{i=1}^{r}[c_i, d_i]$ with $\lceil n/2 \rceil = \min\{\rho(d_i) : i = 1, \ldots, r\}$ in order to show that $\mathrm{sdepth}(\mathfrak{m}) \geq \lceil n/2 \rceil$. Finding such a partition is what Biró et al. succeeded to do.

Let us first show why $\mathrm{sdepth}(\mathfrak{m}) \leq \lceil n/2 \rceil$. For the proof of both inequalities, the easy and the difficult one, we may choose $g = (1, \ldots, 1)$ Then the poset $P_\mathfrak{m}^g$ can be identified with poset P_n of all non-empty subsets $A \subset [n]$ ordered by inclusion, and the function ρ applied to $A \in P_n$ is just the cardinality $|A|$ of A.

We first assume that n is odd, say $n = 2k+1$, and consider an interval partition of $\mathfrak{m}$. Any such partition has to contain intervals of the form $[\{i\}, D_i]$. Since the number of 2-sets in P_n is $(2k+1)(k+1)$, it follows that $\sum_{i=1}^{2k+1} |D_i| = (2k+1)(k+1)$. If the given partition would have Stanley depth $> k + 1$ it would follow that $|D_i| > k + 1$ for all i, contracting the equation $\sum_{i=1}^{2k+1} |D_i| = (2k+1)(k+1)$. Next assume that n is even, say $n = 2k$. Then the number of 2-sets of P_n is $(2k+1)k$. Thus assuming that the given Stanley decomposition has Stanley depth $> k$ would contradict the equation $\sum_{i=1}^{2k+1} |D_i| = (2k + 1)k$.

Now for the proof of the inequality $\mathrm{sdepth}(\mathfrak{m}) \geq \lceil n/2 \rceil$ we may restrict ourselves to the case that $n = 2k + 1$. Indeed, if $\mathscr{P}$ is a partition of P_n with $|D| \geq n/2$ for each interval $[C, D] \in \mathscr{P}$, then $\mathscr{Q} = \mathscr{P} \cup \{[\{n + 1\}, [n + 1]]\}$ is a partition of P_{n+1} with $|D| \geq (n + 1)/2$ for all $[C, D] \in P_{n+1}$.

Now we assume that $n = 2k + 1$. Following [18] we arrange the elements of $[2k + 1]$ around the circle C_n in clockwise star order: the integer i, going clockwise, is followed by the integer j where $j \cong i \bmod 2k + 1$ and $1 \leq j \leq 2k + 1$. For $k = 5$, Fig. 4 displays this order.

Any subset $S \subset [n]$ is the disjoint union of intervals on C_n, called the *blocks* of S, see Fig. 5 where the blocks of $S = \{3, 6, 8, 9, 11\}$ are $\{6, 11\}$ and $\{3, 8, 9\}$ while the gap blocks are $\{1, 2, 7\}$ and $\{4, 5, 10\}$.

A subset $B \subset [n]$ is called *balanced* if each block of $[n] \setminus B$ has even size. The set S in Fig. 5 displays a non-balanced set while the set $B = \{2, 3, 6, 8, 10\}$ is balanced, see Fig. 6. It is clear that the number of elements of each balanced subset of $[n]$ is odd

Fig. 4 Clockwise star order

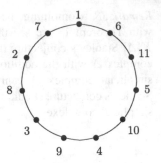

Fig. 5 The blocks for
$S = \{3, 6, 8, 9, 11\}$

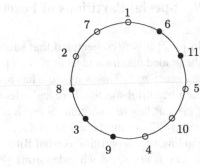

Fig. 6 The balanced set
$B = \{2, 3, 6, 8, 10\}$

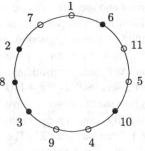

Let B be a balanced subset of $[n]$, say $|B| = 2s + 1$. Let $j \in [n] \setminus B$. Then $|B \cap \{j + 1, \ldots, j + k\}| = s$ or $|B \cap \{j, j + 1, \ldots, j + k\}| = s + 1$, see [18, Lemma 4.2]. The set of elements $j \in [n] \setminus B$ for which the previous intersection contains s elements is denoted L_B, the set of the remaining elements of $[n] \setminus B$ is denoted R_B. In [18, Lemma 4.3] it is shown that $|L_B| = |R_B| = k - s$. It follows that $|B \cup L_B| = k + s + 1$.

For the balanced set $B = \{2, 3, 6, 8, 10\}$ we have $L_B = \{4, 7, 11\}$ and $R_B = \{1, 5, 9\}$.

Now the crucial results is the following

Theorem 39 (Lemma 4.6 in [18]). *Let k be a non-negative integer and let $S \in P_n$ where $n = 2k + 1$. Then there is a unique balanced set B such that $S \in [B, B \cup L_B]$.*

As a consequence we obtain

Corollary 40. *Let k be a non-negative integer and $n = 2k + 1$. Then $\mathscr{P}$: $P_n = \bigcup_{B \text{ is balanced}}[B, B \cup L_B]$ is a partition of the poset $P_n = P_m^g$ with $|B \cup L_B| \geq k + 1$ for all B. In particular, $\mathrm{sdepth}(\mathfrak{m}) \geq \mathrm{sdepth}(\mathscr{D}(\mathscr{P})) \geq k + 1$.*

In the following example we apply Theorems 29 and 39 to obtain a Stanley decomposition $\mathscr{D}$ of $\mathfrak{m} = (x_1, x_2, \ldots, x_5)$ with $\mathrm{sdepth}(\mathscr{D}) = \mathrm{sdepth}(\mathfrak{m}) = 3$. The method of Biró et al. gives us a partition of P_5 with the following intervals:

$$[1, 123], [2, 234], [3, 345], [4, 145], [5, 125],$$

$$[124, 1234], [134, 1345], [135, 1235], [235, 2345], [245, 1245], [12345, 12345].$$

Here $i_1 i_2 \cdots i_r$ stands for $\{i_1, i_2, \ldots, i_r\}$.

The corresponding Stanley decomposition is the following:

$$\begin{aligned}
(x_1, x_2, x_3, x_4, x_5) &= x_1 K[x_1, x_2, x_3] \oplus x_2 K[x_2, x_3, x_4] \oplus x_3 K[x_3, x_4, x_5] \\
&\oplus x_4 K[x_1, x_4, x_5] \oplus x_5 K[x_1, x_2, x_5] \\
&\oplus x_1 x_2 x_4 K[x_1, x_2, x_3, x_4] \oplus x_1 x_3 x_4 K[x_1, x_3, x_4, x_5] \\
&\oplus x_1 x_3 x_5 K[x_1, x_2, x_3, x_5] \oplus x_2 x_3 x_5 K[x_2, x_3, x_4, x_5] \\
&\oplus x_2 x_4 x_5 K[x_1, x_2, x_4, x_5] \\
&\oplus x_1 x_2 x_3 x_4 x_5 x K[x_1, x_2, x_3, x_4, x_5].
\end{aligned}$$

The methods of interval partitions of posets have been extended and refined by Keller et al. [119] to obtain the following results on the Stanley depth of the squarefree Veronese ideal $I_{n,d}$ which is the ideal of all squarefree monomials in n variables generated in degree d.

Theorem 41 (Keller et al. [119]). *Let $I_{n,d}$ be the squarefree Veronese ideal in $S = K[x_1, \ldots, x_n]$, generated by all squarefree monomials of degree d. Then*

(a) *If $1 \leq d \leq n < 5d + 4$, then $\mathrm{sdepth}(I_{n,d}) = \lfloor \frac{n-d}{d+1} \rfloor + d$.*
(b) *If $d \geq 1$ and $n \geq 5d + 4$, then $d + 3 \leq \mathrm{sdepth}(I_{n,d}) \leq \lfloor \frac{n-d}{d+1} \rfloor + d$.*

In general it is conjectured by Cimpoeaş [38] and Keller et al. [119] that $\mathrm{sdepth}(I_{n,d}) = \lfloor \frac{n-d}{d+1} \rfloor + d$ for all positive integers $1 \leq d \leq n$. Some more positive results in this direction have been obtained by Gi et al. [76].

For our next result we will use the following proposition which can be extracted from the work of Keller et al. [119] and Gi et al. [76] and as it is summarized in [62].

Proposition 42. *For each c-set $C \subset [n]$ with $c \leq \lfloor \frac{n+1}{2} \rfloor$, and each non-negative integer s such that $n + 1 \geq (c + 1)(s + 1)$ there exists a superset $C(s)$ of C of cardinality $c + s$ such that the following condition is satisfied: suppose that A and B are subsets of $[n]$ of cardinalities $a \leq b \leq \lfloor \frac{n+1}{2} \rfloor$, and $s' \leq s$ are non-negative integers such that*

$$n + 1 \geq (b + 1)(s' + 1) \geq (a + 1)(s + 1).$$

If B does not belong to $[A, A(s)]$, *then this interval is disjoint from* $[B, B(s')]$.

As an application of this proposition we have the following result [62, Theorem 3.4].

Theorem 43. *Let t be the largest integer such that $n+1 \geq (2t+1)(t+1)$. Then the Stanley depth of any squarefree monomial ideal in a polynomial ring in n variables is greater than or equal to $2t + 1$. In particular, this lower bound is approximately of size $\sqrt{n}$.*

Proof. We consider the sequence $s_1, s_2, \ldots, s_{2t}$ with $s_i = 2t - i + 1$ for $i = 1, \ldots, t$ and $s_i = t$ for $i = t + 1, \ldots, 2s$. Then

$$n + 1 \geq (i + 1)(s_i + 1) \geq i(s_{i-1} + 1) \quad \text{for} \quad i = 2, \ldots, 2t.$$

Now let $I \subset S = K[x_1, \ldots, x_n]$ be a squarefree monomial ideal. Having chosen this sequence we construct a partition $\mathscr{P}$ of the poset P_I^g (with $g = (1, \ldots, 1)$) which we may consider as a subposet of the poset $P(n)$ of all non-empty subsets of $[n]$. For each $\{i\} \in P_I^g$ we take the interval $[\{i\}, \{i\}(s_1)]$. In the next step we take all intervals $[\{i_1, i_2\}, \{i_1, i_2\}(s_2)]$ with $\{i_1, i_2\}$ not contained in any of the intervals $[\{i\}, \{i\}(s_1)]$. In this way we proceed until we reach intervals of the form $[A, A(s_{2t})]$ with $2t$-sets A. By Proposition 42 these intervals are pairwise disjoint and cover the a-sets in P_I^g for all $a \leq 2t$. The remaining set $A \in P_I^g$ which are not covered by the above intervals are all of cardinality $\geq 2t + 1$. Each such a set A we cover by the interval $[A, A]$ to finally obtain the partition $\mathscr{P}$ of the poset P_I^g. Since $A(s_i)$ is a set of cardinality $i + (2t - i + 1) = 2t + 1$ for all our intervals $[A, A(s_i)]$ of $\mathscr{P}$ it follows from Theorem 29 that the corresponding Stanley decomposition has Stanley depth $2t + 1$, so that $\text{sdepth}(I) \geq 2t + 1$.

10 Stanley Decompositions and Alexander Duality

In this section we study the effect of Alexander duality to Stanley decompositions and describe in the special case of Stanley–Reisner rings and Stanley–Reisner ideals the behavior of Stanley decompositions under this duality following the paper [179] where more generally squarefree modules are considered.

Given a simplicial complex Δ on the vertex set $[n]$, we define $\Delta^{\vee}$ by

$$\Delta^{\vee} = \{F^c : F \notin \Delta\}.$$

Here for $F \subset [n]$ we denote by F^c the complement of F in $[n]$, that is, $F^c = [n] \setminus F$.

It is easily checked that $\Delta^{\vee}$ is again a simplicial complex on the vertex set $[n]$. The simplicial complex $\Delta^{\vee}$ is called the *Alexander dual* of Δ. Obviously, one has $\Delta = (\Delta^{\vee})^{\vee}$. More about the Alexander duality can be found in the book [91, Sect. 1.5.3, Chaps. 5.1 and 8.1].

For any collection $\mathscr{A}$ of subsets of $[n]$ we denote by $\mathscr{A}^c$ the set of subsets $\{F^c: F \in \mathscr{A}\}$ of $[n]$.

As an immediate consequence of the definition of the Alexander dual we obtain the following result and its corollary.

Proposition 44. *Let Δ by a simplicial complex. Then $P_{K[\Delta^\vee]} = (P_{I_\Delta})^c$.*

Corollary 45. *Let Δ by a simplicial complex. Then*
$\mathscr{P}$: $P_{K[\Delta^\vee]} = \bigcup_{i=1}^r [F, G]$ *is a partition of $P_{K[\Delta^\vee]}$ if and only if*
$\mathscr{P}^c$: $P_{I_\Delta} = \bigcup_{i=1}^r [G^c, F^c]$ *is a partition of P_{I_Δ}.*

In terms of Stanley decompositions the preceding corollary can be interpreted as follows:

$$\mathscr{D}: K[\Delta^\vee] = \bigoplus_{i=1}^r u_i K[Z_i]$$

is a Stanley decomposition of $K[\Delta^\vee]$ if and only if

$$\mathscr{D}^\vee: I_\Delta = \bigoplus_{i=1}^r v_i K[W_i]$$

is a Stanley decomposition of I_Δ. Here $v_i = \prod_{x_j \in Z_i} x_j$ and $W_i = \{x_j: x_j | u_i\}$.

Let $J \subset I \subset S$ be monomial ideals. For a Stanley decomposition $\mathscr{D}$: $I/J = \bigoplus_{i=1}^r u_i K[Z_i]$ we define the number

$$\mathrm{sreg}(\mathscr{D}) = \max\{\rho(\deg(u_i)): i = 1, \ldots, r\},$$

and set $\mathrm{sreg}(I/J) = \min\{\mathrm{sreg}(\mathscr{D}): \mathscr{D}$ is a Stanley decomposition of $I/J\}$.

The relationship of this invariant to the regularity of a module will be explained in a moment.

We first observe the following fact which can be easily deduced from Corollary 45.

Corollary 46. *Let Δ be a simplicial complex. Then*

$$\mathrm{sreg}(I_\Delta) + \mathrm{sdepth}(K[\Delta^\vee]) = n \quad and \quad \mathrm{sdepth}(I_\Delta) + \mathrm{sreg}(K[\Delta^\vee]) = n.$$

A remarkable generalization of this result due to Okazaki and Yanagawa can be found in [155, Theorem 3.13].

Example 47. Let $\Delta = \{\emptyset\}$. Then $I_\Delta = \mathfrak{m} = (x_1, \ldots, x_n)$ and $K[\Delta^\vee] = S/(x_1 \cdots x_n)$. It follows from Corollary 46 and Proposition 15 that

$$\mathrm{sreg}(\mathfrak{m}) = n - \mathrm{sdepth}(S/(x_1 \cdots x_n)) = n - (n-1) = 1,$$

and from Corollary 46 and the result of Biró et al. that

$$\mathrm{sreg}(S/(x_1,\dots,x_n)) = n - \mathrm{sdepth}(\mathfrak{m}) = n - \lceil n/2 \rceil = \lfloor n/2 \rfloor.$$

Comparing in this example sreg with the Castelnuovo–Mumford regularity, we see that $\mathrm{sreg}(\mathfrak{m}) = \mathrm{reg}(\mathfrak{m})$ and $\mathrm{sreg}(S/(x_1 \cdots x_n)) < \mathrm{reg}(S/(x_1 \cdots x_n)) = n-1$. Is there any relationship between these invariants? To answer this question we recall the following result of Terai, see [91, Proposition 8.1.10]:

$$\mathrm{reg}(I_\Delta) + \mathrm{depth}(K[\Delta^\vee]) = n \text{ and } \mathrm{depth}(I_\Delta) + \mathrm{reg}(K[\Delta^\vee]) = n \qquad (9)$$

Comparing this with the identities in Corollary 46 we see that

$$\mathrm{sreg}(I_\Delta) \le \mathrm{reg}(I_\Delta) \quad \text{and} \quad \mathrm{sreg}(K[\Delta]) \le \mathrm{reg}(K[\Delta]),$$

provided Stanley's conjecture is true. Thus it is natural to conjecture the following inequality.

Conjecture 48 ([179]). Let $J \subset I \subset S$ be monomial ideals. Then $\mathrm{sreg}(I/J) \le \mathrm{reg}(I/J)$.

Actually, Soleyman Jahan conjectured the same inequality for any squarefree $\mathbb{Z}^n$-graded module.

In the following we describe a case where $\mathrm{sreg}(M) \le \mathrm{reg}(M)$. A monomial ideal $I \subset S$ is said to have *linear quotients* if the monomial set of monomial generators of I can be ordered $u_1, u_2, \dots, u_m$ such that for all i the colon ideal $(u_1, \dots, u_{i-1}) : u_i$ is generated by variables. In [91, Proposition 8.2.5] it is shown that I_Δ has linear quotients if and only if $\Delta^\vee$ is shellable.

Proposition 49. *Let I be a squarefree monomial ideal with linear quotients. Then $\mathrm{sreg}(I) = \mathrm{reg}(I)$, and $\mathrm{reg}(I)$ is equal to the maximal degree of an element in the minimal set of monomial generators of I.*

Proof. Let Δ be the simplicial complex with $I = I_\Delta$. Then, by what we said before, the Alexander dual $\Delta^\vee$ is shellable. Therefore it follows from Theorem 13 and Proposition 18 that $\mathrm{sdepth}(S/I_{\Delta^\vee}) = \mathrm{depth}(S/I_{\Delta^\vee})$. Thus the formulas of Terai (9) together with Corollary 46 implies that $\mathrm{sreg}(I) = \mathrm{reg}(I)$. Since I has linear quotients, it follows that I is componentwise linear, see [91, Theorem 8.2.15]. Thus [91, Corollary 8.2.14] completes the proof.

11 The Size of an Ideal

In [129] Lyubeznik introduced the *size* of a monomial ideal and showed that $\mathrm{depth}(S/I) \ge \mathrm{size}(I)$ for any monomial ideal I. In this section we present the result of [97] where it is shown that $\mathrm{sdepth}(I) \ge \mathrm{size}(I) + 1$. Of course, assuming that Stanley's conjecture holds, one expects exactly this inequality.

Let $I \subset S$ be a monomial ideal and $I = \bigcap_{i=1}^{s} Q_i$ a primary decomposition of I, where the Q_i are monomial ideals. Let Q_i be P_i-primary. Then each P_i is a monomial prime ideal and $\mathrm{Ass}(S/I) = \{P_1, \ldots, P_s\}$.

According to [129, Proposition 2] the *size* of I, denoted size(I), is the number $v + (n - h) - 1$, where v is the minimum number t such that there exist $j_1 < \cdots < j_t$ with

$$\sqrt{\sum_{k=1}^{t} Q_{j_k}} = \sqrt{\sum_{j=1}^{s} Q_j},$$

and where $h = \mathrm{height}(\sum_{j=1}^{s} Q_j)$.

Notice that $\sqrt{\sum_{k=1}^{t} Q_{j_k}} = \sum_{k=1}^{t} P_{j_k}$ and $\sqrt{\sum_{j=1}^{s} Q_j} = \sum_{j=1}^{s} P_j$, so that the size of I depends only on the set of associated prime ideals of S/I.

We will indicate the proof of the inequality sdepth$(I) \geq$ size$(I) + 1$. The details of the proof can be found in [97]. For the proof of this inequality the technique of splitting variables is used, as it was introduced in [159, 160]. Let $I \subset S$ be a monomial ideal. We will decompose I into a direct sum of $\mathbb{Z}^n$-graded subspaces. The decomposition depends on the choice of a subset Y of the set of variables $X = \{x_1, \ldots, x_n\}$, and is also determined by the unique irredundant presentation of I as an intersection $I = \bigcap_{j=1}^{s} Q_j$ of its minimal irreducible monomial ideals. As before each Q_j is a P_j-primary ideal with $P_j \in \mathrm{Ass}(S/I)$.

Without loss of generality we may assume that $Y = \{x_1, \ldots, x_r\}$ for some number r such that $0 \leq r \leq n$. Then the set of variables splits into the two sets $\{x_1, \ldots, x_r\}$ and $\{x_{r+1}, \ldots, x_n\}$.

Given a subset $\tau \subset [s]$, we let I_τ be the $\mathbb{Z}^n$-graded K-vector space spanned by the set of monomials of the form $w = uv$ where u and v are monomials with

$$u \in K[x_1, \ldots, x_r] \quad \text{and} \quad u \in \bigcap_{j \notin \tau} Q_j \setminus \sum_{j \in \tau} Q_j,$$

$$v \in K[x_{r+1}, \ldots, x_n] \quad \text{and} \quad v \in \bigcap_{j \in \tau} Q_j.$$

The following result (taken from [97]) extends the corresponding statement shown by Popescu [160] for squarefree monomial ideals.

Proposition 50. *With the notation introduced, the ideal I has a decomposition $\mathscr{D}_Y : I = \bigoplus_{\tau \subseteq [s]} I_\tau$ as a direct sum of $\mathbb{Z}^n$-graded K-subspaces of I.*

Proof. It is clear from the definition of I_τ that $I_\tau \subset I$, so that $\sum_{\tau \subseteq [s]} I_\tau \subset I$. Conversely, let $w = x_1^{a_1} x_2^{a_2} \cdots x_n^{a_n}$ be a monomial in I. Then $w = x_1^{a_1} x_2^{a_2} \cdots x_n^{a_n}$ can be written in a unique way as a product $w = uv$ of monomials with $u \in K[x_1, \ldots, x_r]$ and $v \in K[x_{r+1}, \ldots, x_n]$. Let $\tau = \{j \in [s]: u \notin Q_j\}$. Then $u \in \bigcap_{j \notin \tau} Q_j$ and $u \notin \sum_{j \in \tau} Q_j$.

Let $j \in \tau$. Since $uv \in I$ we have $uv \in Q_j$. Thus, if $Q_j = (x_{i_1}^{b_{i_1}}, \ldots, x_{i_k}^{b_{i_k}})$, then there exists an integer ℓ with $a_{i_\ell} \geq b_{i_\ell}$. On the other hand, since $u \notin Q_j$, it follows that $a_{i_t} < b_{i_t}$ for all $i_t \leq r$. This implies that $i_\ell \geq r + 1$, and consequently $v \in Q_j$. Hence we see that $v \in \bigcap_{j \in \tau} Q_j$, and conclude that $w \in I_\tau$.

In order to see that the sum is direct assume that $w = uv \in I_\tau \cap I_\sigma$. Suppose that $\tau \neq \sigma$. Then we may assume that $\sigma \setminus \tau \neq \emptyset$. Let $j \in \sigma \setminus \tau$. Then $u \in Q_j$ by the definition of I_τ, and $u \notin Q_j$, by the definition of I_σ, a contradiction.

Now we are ready to indicate the proof of the following

Theorem 51. *Let I be a monomial ideal of S. Then*

$$\operatorname{sdepth}(I) \geq \operatorname{size}(I) + 1.$$

Proof. Let $I = \bigcap_{j=1}^s Q_j$ be the unique irredundant presentation of I as an intersection of its minimal irreducible monomial ideals. Each of the Q_j is a primary ideal whose associated monomial prime ideal we denote, as before, by P_j.

We may assume that $\sum_{j=1}^s P_j = \mathfrak{m}$. Indeed, let $Z = \{x_i : x_i \notin \sum_{j=1}^s P_j\}$, $T = K[X \setminus Z]$ and $J = I \cap T$. Then the sum of the associated prime ideals of J is the graded maximal ideal of T, and

$$\operatorname{sdepth}(I) = \operatorname{sdepth}(J) + |Z|, \quad \text{and} \quad \operatorname{size}(I) = \operatorname{size}(J) + |Z|.$$

The first equation follows from [102, Lemma 3.6], while the second equation follows from the definition of size.

We choose the splitting set Y to be the set $\{x_i : x_i \in P_1\}$, and we may assume that $Y = \{x_1, \ldots, x_r\}$ for some number r such that $1 \leq r \leq n$. If $r = n$, then the desired inequality follows at once since in this case $\operatorname{size} I = 0$, and since for every monomial ideal I we have that $\operatorname{sdepth}(I) \geq 1$. Therefore from now on we assume that $r < n$. We will prove the assertion of the theorem by induction on s. The case $s = 1$ follows immediately from [174, Theorem 2.4] and Problem 30 since $\operatorname{sdepth}(Q_1) = \lceil |Y|/2 \rceil + n - |Y|$ and $\operatorname{size} Q_1 = n - |Y|$.

Assume now that the assertion is proved for all monomial ideals which are intersections of at most $s - 1$ irreducible monomial ideals. Since $Y = \{x_1, \ldots, x_r\}$, it follows from the method described before Proposition 50 that $I = \bigoplus_{\tau \subset [s]} I_\tau$ with $I_{[s]} = 0$. We obtain from the decomposition of I that

$$\operatorname{sdepth}(I) \geq \min\{\operatorname{sdepth}(I_\tau) : \tau \subset [s] \text{ and } I_\tau \neq 0\}.$$

The proof is completed by showing that for any subset τ of $[s]$ such that $I_\tau \neq 0$ we have that $\operatorname{sdepth}(I_\tau) \geq \operatorname{size}(I) + 1$. Here the induction hypothesis will be used that the inequality is valid for monomial ideals with at most $s-1$ irreducible components.

Now we consider a concept which in some sense is dual to that of the size of an ideal. Let I be a squarefree monomial ideal minimally generated by the monomials

$u_1, \ldots, u_m$. Let w be the smallest number t with the property that there exist integers $1 \le i_1 < i_2 < \cdots < i_t \le m$ such that

$$\mathrm{lcm}(u_{i_1}, u_{i_2}, \ldots, u_{i_t}) = \mathrm{lcm}(u_1, u_2, \ldots, u_m).$$

Then we call the number $\deg \mathrm{lcm}(u_1, u_2, \ldots, u_m) - w$ the *cosize* of I, denoted cosize I.

Now we have

Proposition 52. *Let $I \subset S$ be a squarefree monomial ideal. Then* $\mathrm{reg}\, S/I \le$ *cosize I.*

Proof. Let Δ be the simplicial complex with the property that $I = I_\Delta$. By using the result of Lyubeznik as well as Terai's result [91, Proposition 8.1.10], we obtain

$$n - \mathrm{reg}\, K[\Delta] = n - \mathrm{proj\, dim}\, I_{\Delta^\vee} = \mathrm{depth}(I_{\Delta^\vee}) \ge \mathrm{size}\, I_{\Delta^\vee} + 1,$$

so that $n - \mathrm{reg}\, K[\Delta] \ge v + (n - h)$. This implies that $\mathrm{reg}\, K[\Delta] \le h - v$.

Let $I_{\Delta^\vee} = P_{F_1} \cap \cdots \cap P_{F_m}$ where the P_{F_i} are the minimal prime ideals of $I_{\Delta^\vee}$. (Here $P_G = (\{x_i\}_{i \in G})$ for $G \subset [n]$.) Then $\{x_{F_1}, \ldots, x_{F_m}\}$ is the minimal monomial set of generators of $I_{\Delta^\vee}$, see [91, Corollary 1.5.5]. (Here $x_G = \prod_{i \in G} x_i$ for $G \subset [n]$.) By using this fact we see that the number v for $I_{\Delta^\vee}$ is equal to the number w for I_Δ, and that the number h for $I_{\Delta^\vee}$ is equal to the number $\deg \mathrm{lcm}(u_1, u_2, \ldots, u_m)$ for I_Δ. This yields the desired result.

Our main result Theorem 51 yields

Corollary 53. *Let $I \subset S$ be a squarefree monomial ideal. Then* $\mathrm{sreg}\, S/I \le$ *cosize I.*

Proof. Let Δ be the simplicial complex with $I = I_\Delta$. Then, by using Corollary 46 as well as Theorem 51, we obtain

$$\mathrm{sreg}\, S/I_\Delta = n - \mathrm{sdepth}(I_{\Delta^\vee}) \le n - (\mathrm{size}\, I_{\Delta^\vee} + 1) = \mathrm{cosize}\, I_\Delta.$$

12 The Hilbert Depth

Let M be a $\mathbb{Z}^n$-graded S-module and $\mathscr{D}: M = \bigoplus_{i=1}^r m_i K[Z_i]$ a Stanley decomposition. As in the proof of Proposition 33 this decomposition yields the following formula for the Hilbert series of M (viewed as a $\mathbb{Z}$-graded module over the standard graded polynomial ring S):

$$\mathscr{H}(\mathscr{D}): \mathrm{Hilb}(M) = \sum_{i=1}^{r} \frac{t^{a_i}}{(1-t)^{b_i}}, \tag{10}$$

where $a_i = \deg(m_i)$ and $b_i = |Z_i|$.

Let $\mathcal{H}$ be a sum presentation of Hilb(M) as in (10) (not necessarily induced by a Stanley decomposition). We call such a sum presentation a *Hilbert decomposition* of Hilb(M), and call the number

$$\text{hdepth}(\mathcal{H}) = \min\{b_i\colon i = 1,\ldots,r\}$$

the *Hilbert depth* of $\mathcal{H}$. Finally we define the *Hilbert depth* of M to be the number

$$\text{hdepth}(M) = \max\{\text{hdepth}(\mathcal{H})\colon \mathcal{H} \text{ is a Hilbert decomposition of Hilb}(M)\}.$$

This invariant has been introduced in [31] where it is denoted $\text{Hdepth}_1(M)$. In that paper one can find a multigraded version of this definition as well, which the authors denote by $\text{Hdepth}_n(M)$.

It is obvious from the definition of the Hilbert depth that the following inequality holds

$$\text{hdepth}(M) \geq \text{sdepth}(M). \tag{11}$$

Thus one would expect that $\text{hdepth}(M) \geq \text{depth}(M)$. That this is indeed the case was shown in [31, Theorem 2.7]. Here we prove this result for the special case $M = S/I$, by using a different argument.

Theorem 54. *Let $I \subset S$ be a monomial ideal. Then*

$$\text{hdepth}(S/I) \geq \text{depth}(S/I).$$

Proof. Let $\text{gin}(I)$ be the generic initial ideal of I with respect to the reverse lexicographic order. By Galligo, Bayer and Stillman (see [91, Theorem 4.2.1] and [91, Theorem 4.2.10]) it is known that $\text{gin}(I)$ is of Borel type. Thus it follows from Proposition 21 that $S/\text{gin}(I)$ is pretty clean which by Proposition 18 implies that $\text{sdepth}(S/\text{gin}(I)) = \text{depth}(S/\text{gin}(I)) = \text{depth}(S/I)$, see [91, Corollary 4.3.18]. Thus there exists a Stanley decomposition of $S/\text{gin}(I)$ such that Hilbert depth of $\mathcal{H}(\mathcal{D})$ is equal to $\text{depth}(S/I)$. Hence, since $\text{Hilb}(S/I) = \text{Hilb}(S/\text{in}(I))$, we see that $\text{hdepth}(S/I) \geq \text{depth}(S/I)$.

As an example for the computation of the Hilbert depth consider the maximal ideal $\mathfrak{m} = (x_1,\ldots,x_n)$ in S. Let

$$\mathcal{H}\colon \text{Hilb}(\mathfrak{m}) = \sum_{i=1}^{r} \frac{t}{(1-t)^{b_i}} + \sum_{i=1}^{s} \frac{t^2}{(1-t)^{c_i}} + \cdots = \sum_{i=1}^{r} t(1+b_i t+\cdots)+st^2+\cdots$$

be a Hilbert decomposition of Hilb($\mathfrak{m}$). Since $\text{Hilb}(\mathfrak{m}) = nt + \binom{n+1}{2}t^2 + \cdots$, comparison of coefficients gives $r = n$ and $\sum_{i=1}^{n} b_i \leq \binom{n}{2}$. This implies that $\min\{b_i\colon i = 1,\ldots,n\} \leq (n + 1)/2$. It follows that $\text{hdepth}(\mathcal{H}) \leq \lfloor (n + 1)/2 \rfloor = \lceil n/2 \rceil$ Therefore $\text{hdepth}(\mathfrak{m}) \leq \lceil n/2 \rceil = \text{sdepth}(\mathfrak{m})$, see Sect. 9.

Since hdepth(M) $\geq$ sdepth(M) for any $\mathbb{Z}^n$-graded S-module, we finally get that hdepth(m) $= \lceil n/2 \rceil$.

Comparing Formula (10) with the definition of the Stanley regularity as defined in Sect. 10 it is natural to define the Hilbert regularity of a Hilbert decomposition $\mathscr{H}$: Hilb(M) $= \sum_{i=1}^{r} \frac{t^{a_i}}{(1-t)^{b_i}}$ of a $\mathbb{Z}^n$-graded module M to be the number hreg(M) $= \max\{a_i : i = 1, \ldots, r\}$, and to define the *Hilbert regularity* of M as the number

$$\text{hreg}(M) = \min\{\text{hreg}(\mathscr{H}) : \mathscr{H} \text{ is a Hilbert decomposition of Hilb}(M)\}.$$

Obviously one has that hreg(M) $\leq$ sreg(M), but also

Theorem 55. *Let M be a $\mathbb{Z}^n$-graded S-module. Then* hreg(M) $\leq$ reg(M).

Proof. We prove the assertion by induction on dim(M). If dim(M) $= 0$, then Hilb(M) is a polynomial of degree reg(M). Therefore, in this case, hreg(M) $=$ reg(M). Now assume that dim(M) > 0, and let $N \subset M$ be the maximal submodule of M with finite length. Then we obtain the exact sequence

$$0 \to N \longrightarrow M \overset{}{\longrightarrow} W \longrightarrow 0,$$

where W is a module of positive depth. By the definition of Hilbert regularity we get

$$\text{hreg}(M) \leq \max\{\text{hreg}(N), \text{hreg}(W)\} = \max\{\text{reg}(N), \text{hreg}(W)\}. \qquad (12)$$

Without restriction we may assume that the base field of S is infinite. Then there exists a nonzero divisor $y \in S$ on W of degree 1. Let $\sum_{i=1}^{r} t^{a_i}/(1-t)^{b_i}$ be a Hilbert decomposition of $H_{W/yW}(t)$, then $\sum_{i=1}^{r} t^{a_i}/(1-t)^{b_i+1}$ is a Hilbert decomposition of W. This implies that hreg(W) $\leq$ hreg(W/yW). Thus applying our induction hypothesis we see that

$$\text{hreg}(W) \leq \text{hreg}(W/yW) \leq \text{reg}(W/yW) = \text{reg}(W).$$

Therefore (12) implies that hreg(M) $\leq \max\{\text{reg}(N), \text{reg}(W)\} = \text{reg}(M)$, as desired. For the last equation we used [52, Corollary 20.19].

Problem 56. Find an example of a $\mathbb{Z}^n$-graded S-module M such that hdepth(M) $>$ sdepth(M).

Problem 57. Given a $\mathbb{Z}^n$-graded module M with Hilbert decomposition as in (10). Show that $b_i \leq d = \dim(M)$ for all i and that the number of elements i with $b_i = d$ is equal to the multiplicity of the module M.

13   Further Results and Open Problems

In this section we summarize what is known about Stanley decompositions and related topics which have not been discussed in the previous sections.

1. *What is know about Stanley's conjecture?*

Apel in his papers [12, 13] was the first to prove Stanley's conjecture in a several interesting special cases. In his first paper [12] he proved Stanley's conjecture for generic monomial ideals. Precisely his result is the following:
([12, Theorem 2]) Let $I \subset S$ be a monomial ideal. Furthermore, assume that for some variable x_k the ideal I has the following property: for any two distinct minimal generators m_i and m_j of I such that $\deg_{x_k}(m_i) = \deg_{x_k}(m_j) = d$ there exists is a third minimal generator m_r which divides $\mathrm{lcm}(m_i, m_j)$ and satisfies $\deg_{x_k} m_r < d$. Then $\mathrm{sdepth}(I) \geq \mathrm{depth}(I)$.

Here Apel used a refinement (see [142]) of the original definition [16] of generic monomial ideals. In the same paper he proved
([12, Theorem 1]) Stanley's conjecture holds true for any monomial ideal in a polynomial ring with at most three variables.

These results have been completed by Apel in the second paper [13] as follows: in Corollary 2 he shows (with a different argument as presented here in Sect. 4) that S/I satisfies Stanley's conjecture for any Borel-fixed ideal, and in Corollary 3 he shows that S/I satisfies Stanley's conjecture whenever $\dim(S/I) \leq 1$. Moreover, he showed
([13, Theorem 3]) Let I be a generic monomial ideal. Then

$$\mathrm{sdepth}(S/I) = \min\{\mathrm{depth}(S/P): P \in \mathrm{Ass}(S/I)\}.$$

In particular, together with Theorem 9, it follows that $\mathrm{sdepth}(S/I) \geq \dim(S/I)$ if I is a generic monomial ideal. In [13, Theorem 5] Apel also showed that S/I satisfies Stanley's conjecture if S/I is Cohen–Macaulay and I is a cogeneric monomial ideal.

According to Sturmfels [184] a monomial ideal I with the irredundant irreducible decomposition $I = \bigcap_{i=1}^{m} \mathfrak{m}^{a_i}$ is said to be *cogeneric*, if any distinct $\mathfrak{m}^{a_i}$ and $\mathfrak{m}^{a_j}$ do not have the same minimal monomial generators. Here, for $a \in \mathbb{N}$, $\mathfrak{m}^a$ denotes the irreducible ideal $(x_i^{a_i}: a_i > 0)$. Okazaki and Yanagawa improved Apel's result as follows:
([155, Theorem 6.5]) If I is a cogeneric monomial ideal, then $\mathrm{sdepth}(S/I) \geq \mathrm{depth}(S/I)$. That is, Stanley's conjecture holds for the quotient by a cogeneric monomial ideal (no matter whether S/I is Cohen–Macaulay or not).

Finally I want to mention the following result of Apel:
([13, Theorem 4]) Let $I \subset S = K[x_1, x_2, x_3]$ be a monomial ideal. Then $\mathrm{sdepth}(S/I) = \min\{\mathrm{depth}(S/P): P \in \mathrm{Ass}(S/I)\}$. In particular, Stanley's conjecture holds for S/I, when S is a polynomial ring in at most three variables.

The most far reaching result in this direction is that of Popescu who showed in [161, Theorem 4.3] that S/I satisfies Stanley's conjecture for a polynomial ring in at most five variables. Anwar and Popescu [11] also give an affirmative answer to Stanley's conjecture when I has at most three irreducible components:
([162, Theorems 5.6 and 5.9]) Stanley's conjecture holds true for $Q_1 \cap Q_2$ and for $S/(Q_1 \cap Q_2 \cap Q_3)$ where Q_1, Q_2, Q_3 are nonzero irreducible monomial ideals of S.

In [37, Theorem 2.3] Cimpoeaş showed that Stanley's conjecture holds for I and S/I when I is a monomial ideal generated by at most three elements.
For monomial ideals of low codimension the following is known by Soleyman Jahan, Yassemi and myself:
([98, Proposition 1.4 and Theorem 2.1]) Let I be a monomial ideal which is perfect of codimension 2 or Gorenstein of codimension 3. Then S/I satisfies Stanley's conjecture.

Another case of interest proved by Soleyman Jahan in the same paper is the following:
([178, Proposition 2.1]) Let $I \subset S = K[x_1, \ldots, x_n]$ be a monomial ideal of height $\geq n - 1$, then Stanley's conjecture holds for S/I.
In a very recent paper [14], Bandari, Divaani-Aazar and Soleyman Jahan showed that if $I \subset S = K[x_1, \ldots, x_n]$ is a monomial ideal generated by monomials $u_1, u_2, \ldots, u_t$, then S/I is pretty clean (and hence satisfies Stanley's conjecture) if either: $u_1, u_2, \ldots, u_t$ is a filter-regular sequence, or d-sequence, or I is an almost complete intersection.
Finally we would like to mention the following result due to Pournaki, Fakhari and Yassemi.
([163, Corollary 2.8] Let I be the edge ideal of a graph. If G is a forest, then S/I^k satisfies Stanley's conjecture for all $k \gg 0$.

2. Numerical bounds for the Stanley depth in special cases.

The following results in this subsection are obtained by using Theorem 29 and extensions of the techniques of constructing partitions as developed by Biró, Howard, Keller, Trotter and Young in [18], and as outlined in Sect. 9.
We first would like to mention the following formula by Shen which is a natural extension of the result in [18] where it was shown that $\text{sdepth}(\mathfrak{m}) = \lceil n/2 \rceil$ for the graded maximal ideal of $S = K[x_1, \ldots, x_n]$.
([174, Theorem 2.4]) Let $I \subset K[x_1, \ldots, x_n]$ be a complete intersection monomial ideal minimally generated by m elements. Then $\text{sdepth}(I) = n - \lfloor m/2 \rfloor$.
This result has then been complemented by Okazaki who obtain the following nice lower bound for the Stanley depth of a monomial ideal
([154, Theorem 2.1]) Let $I \subset K[x_1, \ldots, x_n]$ be a monomial ideal minimally generated by m monomials. Then $\text{sdepth}(I) \geq n - \lfloor m/2 \rfloor$.
The squarefree version of this inequality by Okazaki was first proved by Keller and Young [120, Theorem 1.1].
Another extension of the result of Biró et al. concerns the so-called squarefree Veronese ideals $I_{n,d}$. The ideal $I_{n,d}$ is the ideal of all squarefree monomials in $K[x_1, \ldots, x_n]$ of degree d. It has been conjectured by Cimpoeaş

[38, Conjecture 1.6] and by Keller et al. [119, Conjecture 2.4] that $\mathrm{sdepth}(I_{n,d}) = \lfloor\binom{n}{d+1}/\binom{n}{d}\rfloor + d$. Cimpoeaş [38] as well as Keller et al. [119] showed that the above conjectured Stanley depth is certainly an upper bound for $\mathrm{sdepth}(I_{n,d})$. The most far reaching result in this direction is the following:

([76, Theorem 1.2]) One has $\mathrm{sdepth}(I_{n,d}) = \lfloor\binom{n}{d+1}/\binom{n}{d}\rfloor + d$ for $1 \le d \le n \le (d+1)\lfloor(1+\sqrt{5+4d})/2\rfloor + 2d$.

3. Further results on the Hilbert depth.

In Sect. 12 we have seen that $\mathrm{hdepth}(\mathfrak{m}) = \mathrm{sdepth}(\mathfrak{m}) = \lceil n/2\rceil$ where $\mathfrak{m}$ denotes the graded maximal ideal of $K[x_1,\ldots,x_n]$. Bruns, Krattenthaler and Uliczka generalized this result and showed

([32, Theorem 1.2]) For all n and k one has $\mathrm{hdepth}(\mathfrak{m}^k) = \lceil n/(k+1)\rceil$.

The proof of this result is surprisingly involved and needs a careful numerical analysis. It is conjectured by Cimpoeaş [36] that, similarly as for $\mathfrak{m}$ one has that $\mathrm{hdepth}(\mathfrak{m}^k) = \mathrm{sdepth}(\mathfrak{m}^k)$ for all k. Cimpoeaş actually showed in [36, Theorem 2.2] that $\mathrm{sdepth}(\mathfrak{m}^k) \le \lceil n/(k+1)\rceil$. Of course this result is also a consequence of the above quoted result of Bruns et al.

In [31] Bruns, Krattenthaler and Uliczka study the Stanley depth of the Koszul cycles, that is, of the syzygy modules of $S/\mathfrak{m}$. In their paper the kth syzygy module of $S/\mathfrak{m}$ is denoted $M(n,k)$. It is shown

([31, Theorem 3.5 and Proposition 2.6] $\mathrm{hdepth}(M(n,k)) = n-1$ for $n > k \ge \lfloor n/2\rfloor$, and $\mathrm{hdepth}(M(n,k)) \le n - \lceil(n-k)/(k+1)\rceil$, if $k < \lfloor n/2\rfloor$.

The precise Hilbert depth in the lower range for k is not known.

4. Auxiliary results

Let $I \subset S$ be a monomial ideal. We denote by $\sqrt{I}$ the radical of I. Is the any comparison between the Stanley depth of S/I and $S/\sqrt{I}$. The following results are known. Apel showed

([13, Theorem 1]) $\mathrm{sdepth}(S/\sqrt{I}) \ge \mathrm{sdepth}(S/I)$.

This result has been extended by Ishaq as follows:

([111, Theorem 2.1]) Let $J \subset I \subset S$ be monomial ideals:

then $\mathrm{sdepth}(\sqrt{I}/\sqrt{J}) \ge \mathrm{sdepth}(I/J)$.

Ishaq also proved the following interesting upper bound for the Stanley depth of a monomial ideal.

([112, Theorem 1.1]) Let $I \subset S$ be a monomial ideal with $\mathrm{Ass}(S/I) = \{P_1,\ldots,P_s\}$. Then $\mathrm{sdepth}(I) \le \min\{\mathrm{sdepth}(P_i): i = 1,\ldots,s\}$.

A remarkable lower bound for the Stanley depth has been found by S.A. Seyed Fakhari:

([172, Corollary 3.4]) Let $I \subset K[x_1,\ldots,x_n]$ be a squarefree monomial ideal which is generated in a single degree. Then $\mathrm{sdepth}(I) \ge n - l(I) + 1$ and $\mathrm{sdepth}(S/I) \ge n - l(I)$, where $l(I)$ denotes the analytic spread of I.

One may ask how Stanley depth behaves with respect to monomial localization. Let $I \subset S = K[x_1,\ldots,x_n]$ be a monomial ideal. Monomial localization of I with respect to the variable x_n is the ideal $J \subset S' = K[x_1,\ldots,x_{n-1}]$ which is obtain from I by the homomorphism which maps x_n to 1 and leaves the other variables unchanged. Note that $I_{x_n} = JS'[x_n, x_n^{-1}]$, where I_{x_n} denotes the usual localization of I with respect to x_n. Nasir showed that Stanley depth behaves as follows with respect to monomial localization.

([148, Corollary 3.2]) Let $I \subset S = K[x_1, \ldots, x_n]$ be a monomial ideal and $J \subset S' = K[x_1, \ldots, x_{n-1}]$ the monomial localization of I with respect to x_n. Then $\mathrm{sdepth}(S/I) \geq \mathrm{sdepth}(S'/J) - 1$.

Let $f = \prod_{j=1,\ldots,k} x_{i_j}$. Then $S_f = S[x_{i_1}^{-1}, \ldots, x_{i_k}^{-1}]$. In [149] Nasir and Rauf defined Stanley decomposition and Stanley depth for ideals in the ring S_f, and showed that the number o maximal Stanley spaces in any Stanley decomposition of S_f is equal to 2^k.

In [136] MacLagan and Smith describe an algorithm to obtain a Stanley decomposition of S/I. A special case of this algorithm occurs implicitly already in the proof of [186, Lemma 2.4].

([136, Algorithm 3.4]) If I is (monomial) prime ideal, then $S/I = K[Z]$ where Z is the set of variables not belonging to I. Now assume I is not a prime ideal, and choose a variable x_i that is a proper divisor of a minimal generator of I. One obtains the exact sequence

$$0 \longrightarrow (S/(I : x_i))(-\epsilon_i) \xrightarrow{\ x_i\ } S/I \longrightarrow S/(I + x_i) \longrightarrow 0.$$

Here ϵ_i is the canonical ith unit vector in $\mathbb{Z}^n$.

By Noetherian induction we may assume that we have a Stanley decomposition $\mathscr{D}_1 : (S/(I : x_i)) = \bigoplus_{j=1}^r u_j K[Z_j]$, and a Stanley decomposition $\mathscr{D}_2 : S/(I + x_i) = \bigoplus_{k=1}^s v_k K[W_k]$. Then

$$\mathscr{D} : S/I = \bigoplus_{j=1}^r x_i u_j K[Z_j] \oplus \bigoplus_{k=1}^s v_k K[W_k]$$

is a Stanley decomposition of S/I.

5. Some open problems.

Let $S = K[x_1, \ldots, x_n]$ be the polynomial over K in n-indeterminates. In the previous subsection we mentioned the conjecture by Cimpoeaş according to which one should have $\mathrm{sdepth}(\mathfrak{m}^k) = \lceil n/(k+1) \rceil$ for all n and k where $\mathfrak{m}$ is the graded maximal ideal of S. The squarefree version of this conjecture is the following: $\mathrm{sdepth}(I_{n,d}) = \lfloor \binom{n}{d+1}/\binom{n}{d} \rfloor + d$. It would be a challenge to settle these conjectures by finding suitable partitions of the attached characteristic posets.

Assuming Cimpoeaş's conjecture one has $\mathrm{sdepth}(\mathfrak{m}^k) \geq \mathrm{sdepth}(\mathfrak{m}^{k+1})$ for all k.

Question 58. Let $I \subset S$ be a monomial ideal. Is it true that $\mathrm{sdepth}(I^k) \geq \mathrm{sdepth}(I^{k+1})$ for all k? In general such an inequality is not true for the ordinary depth. There exist examples with $\mathrm{depth}(I^k) < \mathrm{depth}(I^{k+1})$, see [89, 145]. In general this inequality is also not true for the Stanley depth. Consider for example the monomial ideal $I = (x_1^4, x_1^3 x_2, x_1 x_2^3, x_2^4, x_1^2 x_2^2 x_3)$. Then $\text{depth}(S/I) = 0$ and $\text{depth}(S/I^2) = 1$. It follows from Theorem 27 that $\text{sdepth}(S/I) = 0$ and $\text{sdepth}(S/I^2) > 0$.

On the other hand, S.A. Seyed Fakhari [171] showed that for each monomial ideal I and integer $k \geq 1$ the inequalities $\text{sdepth}(S/\overline{I}) \geq \text{sdepth}(S/\overline{I^k})$ and $\text{sdepth}(\overline{I}) \geq \text{sdepth}(\overline{I^k})$ hold. Here $\overline{J}$ denotes the integral closure of the ideal J. Thus Question 58 may have a positive answer for normal monomial ideals, that is, for monomial ideals for which all of its powers are integrally closed.

It is known by Brodmann [28] that for any graded ideal $I \subset S$ there exists an integer such that $\text{depth}(I^k) = \text{depth}(I^{k+1})$ for all $k \geq k_0$.

Conjecture 59. Let $I \subset S$ be a monomial ideal. Then there exists an integer k_0 such that $\text{sdepth}(I^k) = \text{sdepth}(I^{k+1})$ for all $k \geq k_0$. A similar statement holds for the Hilbert depth.

For the maximal ideal m we have $\text{sdepth}(\text{m}^k) = 1$ for all $k \geq n$, by [36, Theorem 2.2] or [32, Theorem 1.2].

Conjecture 60. Let $I \subset S$ be a monomial ideal. Then there exists an integer k_1 such that $\text{sdepth}(\text{m}^k I) = \text{hdepth}(\text{m}^k I) = 1$ for all $k \geq k_1$.

For the proof of this conjecture it suffices to show that $\text{hdepth}(\text{m}^k I) = 1$ for all $k \geq k_1$, because $1 \leq \text{sdepth}(\text{m}^k I) \leq \text{hdepth}(\text{m}^k I)$ for all k.

It is known that for any graded ideal $I \subset S$, the regularity $\text{reg}(I^k)$ is a linear function of k for $k \gg 0$.

Conjecture 61. Let $I \subset S$ be a monomial ideal. Then there exists integers k_2, a and b such that $\text{sreg}(I^k) = ak + b$ for $k \geq k_2$.

Let $I \subset S$ be a monomial ideal. In [178, Theorem 3.10] Soleyman Jahan showed that S/I is pretty clean if and only if S/I^p is clean. Here I^p denotes the polarization of I, see [30, Lemma 4.2.16]. Consequently, if S/I is pretty clean, then not only S/I but also S/I^p satisfy Stanley's conjecture.

Conjecture 62. Let $I \subset S$ be a monomial ideal. Then

$$\text{sdepth}(S/I) - \text{depth}(S/I) = \text{sdepth}(S^p/I^p) - \text{depth}(S^p/I^p).$$

Here S^p is the polynomial ring where I^p is defined.

The conjecture in combination with Corollary 37 implies that Stanley's conjecture holds for all K-algebras with monomial relations if and only if any Cohen–Macaulay simplicial complex is partionable.

For the proof of this conjecture it suffices to consider a 1-step polarization. The complete polarization is obtained by a sequence of 1-step polarizations. A 1-step polarization is obtained as follows: Let $I = (u_1, \ldots, u_m)$ be the minimal set of monomial generators of I. Fix a number $i \in [n]$. We define the 1-step polarization of I with respect to i to be the monomial ideal $J = (v_1, \ldots, v_m) \subset S[y]$, where y is an indeterminate over S and

$$v_j = \begin{cases} y(u_j/x_i), & \text{if } x_i^2 \text{ divides } u_j, \\ u_j, & \text{otherwise.} \end{cases}$$

It is known that $\text{depth}(S[y]/J) = \text{depth}(S/I) + 1$. Thus for the proof of Conjecture 62 one has to show that $\text{sdepth}(S[y]/J) = \text{sdepth}(S/I) + 1$.

Let M be a finitely generated $\mathbb{Z}^n$-graded S-module with syzygy module Z_k for $k = 1, 2, \ldots$. We have seen in Theorem 28 that $\text{sdepth}(Z_k) \geq k$.

Question 63. Is it true that $\text{sdepth}(Z_{k+1}) \geq \text{sdepth}(Z_k)$?

In [31, Lemma 3.2] Bruns et al. showed that this question has a positive answer for the syzygies of the graded maximal ideal of S. On the other hand, even the following conjecture (which is a very special case of the preceding question) is widely open.

Conjecture 64. Let $I \subset S$ be a monomial ideal.
Then $\text{sdepth}(I) \geq \text{sdepth}(S/I)$.

In all known cases one even has $\text{sdepth}(I) > \text{sdepth}(S/I)$.

Question 65. Does there exist an algorithm to compute the Stanley depth of finitely generated $\mathbb{Z}^n$-graded S-modules?

The following problem has a good chance to be solved.

Problem 66. Find an algorithm to compute the Stanley depth for finitely generated graded S-modules M with $\dim_K M_a \leq 1$ for all a.

However if we drop the assumption that $\dim_K M_a \leq 1$ for all a, an answer to Question 65 seems to be hard. The next problem demonstrates how little is known.

Problem 67. Let M and N be finitely generated graded S-modules. Then

$$\text{sdepth}(M \oplus N) \geq \min\{\text{sdepth}(M), \text{sdepth}(N)\}.$$

Do we have equality?

Let $I \subset S$ be a monomial ideal. If equality holds in Problem 67, then in particular one has $\text{sdepth}(I \oplus S) = \text{sdepth}(I)$. To the best of my knowledge this is not known.

Stanley Decompositions Using CoCoA

Anna Maria Bigatti and Emanuela De Negri

1 First Steps with CoCoA-5

First released in 1988, CoCoA is a *special-purpose* Computer Algebra System for doing **C**omputations in **C**ommutative **A**lgebra. It is *freely available* and offers a textual interface, an Emacs mode, and a graphical user interface common to most platforms [39].

Lately is has been entirely rewritten and now comprises three main components: a C++ library (CoCoALib), an interactive system (CoCoA-5), and an algebra computation server (CoCoAServer). Of these components CoCoALib is the heart; it embodies all the "mathematical knowledge" and it is the most evolved part [1]. The roles of the other two parts are to make CoCoALib's capabilities more readily accessible.

Users of CoCoA-4 can move easily to CoCoA-5 because the new CoCoALanguage is highly compatible with the old one; moreover error handling has been greatly improved making it even simpler to use (and to adapt to the new rules).

For both old and new users we start now from the very beginning.

The first step consists of defining the ring in which we want to work, for example $R = \mathbb{Q}[x_1, \ldots, x_4]$.

A.M. Bigatti (✉) · E. De Negri
Dipartimento di Matematica Università degli Studi di Genova Via Dodecaneso 35,
16146 Genova, Italy
e-mail: bigatti@dima.unige.it; denegri@dima.unige.it

A.M. Bigatti et al. (eds.), *Monomial Ideals, Computations and Applications*,
Lecture Notes in Mathematics 2083, DOI 10.1007/978-3-642-38742-5_2,
© Springer-Verlag Berlin Heidelberg 2013

```
Use R ::= QQ[x[1..4]];
```

The special assignment symbol ": : =" (instead of " : =") indicates that what follows is to be interpreted as a polynomial ring (normally " [ ] " indicates getting a component of a list).

If an expression *expr* is assigned then it is not printed, otherwise it is printed: equivalent to "Print *expr*".

The multiplication sign "*" is mandatory, but it can be omitted in expressions between "***" using single lower-case letters (with or without indices) as polynomial indeterminates:

```
M := x[1]^2*x[2]*x[4]^2;
P := *** x[1]x[2]x[3] ***;
```

Text following "--" is a comment, and we will use this to describe our code

```
I := ideal(M, P);     -- the Monomial Ideal generated by M and P
I;  -- print I

Log(M); -- list of the exponents of M
Log(P);

-- viceversa: LogToTerm
L := [0,1,2,1];
LogToTerm(R,L);  -- NEW in CoCoA-5: we need to specify the ring

-- NB:  "="  is for equality test and  ":="  is for assignment
LogToTerm(R, Log(P)) = P;
Log(LogToTerm(R,L)) = L;

-- How to convert a list of indices into a squarefree monomial:
L := [1,2,3];
M := Product([ x[i] | i In L ]);
M;  --> will print  x[1]*x[2]*x[3]
-- ...and back:
LogM := Log(M);
[ i In 1..4 | LogM[i]>0 ];    -- 1..4 is the list [1,2,3,4]
```

(For CoCoA-4 users: lower case letters, such as "i", can now be used in CoCoA-5 also for *programming variables*)

To convert this code into a function just write it between "Define ...EndDefine" specifying the *name* and the *arguments* of the function, and which expression you want to *return*:

```
Define Indices(M)
  LogM := Log(M);
  Return [ i In 1..Len(LogM) | LogM[i]>0 ];
EndDefine
```

...and then try it:

```
Indices(P);   --> will print  [1,2,3]
```

2 Stanley–Reisner Ideals

In this section we deal with simplicial complexes. In particular we write procedures to pass from a simplicial complex to its Stanley–Reisner ideal and back, by using the Alexander dual. For all the definitions see the end of Sect. 10 in chapter "A Survey on Stanley Depth".

AlexanderDual and PrimaryDecomposition are already implemented in CoCoA, but here we see them as an example of CoCoALanguage programming.

Recall that if $I \subseteq K[x_1, \ldots, x_n]$ is a squarefree monomial ideal, then

$$I = \bigcap_{j=1}^{r} P_{F_j},$$

where for a subset $F \subset [n]$ we set $P_F = (\{x_i \; : \; i \in F\})$. If moreover we set $x_F = \prod_{i \in F} x_i$, then the ideal

$$I^\vee = (x_{F_1}, \ldots, x_{F_r})$$

is called the Alexander dual of I. By duality one has that if $I = (m_1, \ldots, m_t)$, then

$$I^\vee = \bigcap_{i=1}^{t} P_{m_i},$$

where $P_m = (\{x_i \; : \; i \in \operatorname{supp}(m)\})$. Here for a monomial m we set $\operatorname{supp}(m) = \{k \; : \; x_k | m\}$.

```
Define IndetsIdeal(M) -- with M a monomial, returns the ideal P_M
  L := Log(M);
  P := RingOf(M);
  Return Ideal([Indet(P,i) | i In 1..NumIndets(P) And L[i]<>0]);
EndDefine; -- IndetsIdeal

IndetsIdeal(x[1]*x[2]);
---->  ideal(x[1], x[2])

Define MyAlexanderDual(I)
  Return IntersectionList([IndetsIdeal(G) | G In Gens(I)]);
EndDefine; -- SqFrAlexanderDual
```

Since $(I^\vee)^\vee = I$, we get the primary decomposition of I by calculating the Alexander dual of I.

```
Define MyPrimaryDecomposition(I)
  AD := MyAlexanderDual(I);
  Return [IndetsIdeal(G) | G In Gens(AD)];
EndDefine; -- SqFrPrimaryDecomposition
```

```
MyAlexanderDual(Ideal(x[1]*x[2], x[2]*x[3])));
---->  ideal(x[2], x[1]*x[3])
```

To make our code more robust we add some input checking:

```
Define IndetsIdeal(M)
  If Not IsTerm(M) Then Error("not a term!"); EndIf;
  L := Log(M);
  If Max(L)>1 Then Error("not squarefree"); EndIf;
  P := RingOf(M);
  Return Ideal([Indet(P,i) | i In 1..NumIndets(P) And L[i]<>0]);
EndDefine; -- IndetsIdeal
```

Now we come back to simplicial complexes.

Let Δ be a simplicial complex on the vertex set $\{1, 2, \ldots, n\}$ and let $I_\Delta \subseteq K[x_1, \ldots, x_n]$ be the Stanley–Reisner ideal of Δ, that is, $I_\Delta = (x_{i_1} x_{i_2} \cdots x_{i_t} \mid \{i_1, i_2, \ldots, i_t\} \notin \Delta)$

Recall that:

1. Δ is determined by the set $\mathscr{F}(\Delta)$ of its facets (maximal faces).
2. The minimal prime ideals of I_Δ are of the form $(x_{j_1}, x_{j_2}, \ldots, x_{j_s})$ with $\{j_1, \ldots, j_s\}$ complement of a facet.
3. The Alexander dual of I_Δ is generated by all the monomials $x_{j_1} x_{j_2} \cdots x_{j_s}$ with $\{j_1, \ldots, j_s\}$ complement of a facet.

We encode the simplicial complex Δ with vertices $\{1, 2, 3, 4, 5\}$ as a list of facets. In the following we consider a complex Δ and we calculate the Stanley Reisner ideal I_Δ.

```
Use QQ[x[1..5]];
-- Delta: Facets of the complex
Delta := [[1,2,3], [2,3,4], [1,4,5]];
-- CDelta: complements of the Facets
CDelta := [ Diff(1..5,F) | F In Delta ];  CDelta;
---->  [[4, 5], [1, 5], [2, 3]]

-- ideal of the indets indexed in CDelta[1]
Ideal([x[j] | j In CDelta[1]]);
---->  ideal(x[4], x[5])

-- list of all the ideals
L := [ Ideal([x[j] | j In C]) | C In CDelta ];

-- intersection of all the elements in L
IDelta := IntersectionList(L);
IDelta; -- Stanley Reisner ideal of Delta
---->  ideal(x[3]*x[5], x[2]*x[5],
----          x[1]*x[3]*x[4], x[1]*x[2]*x[4])
```

We can convert this code into a function in two ways, more direct (importing the symbol "x" from the top-level environment) or more robust (passing the ring "R" and using "Indets(R)"):

```
-- import x from top-level
Define StanleyReisnerIdeal(D)
  TopLevel x;
  CD := [Diff(1..5,F) | F In D];
  SR := IntersectionList ([Ideal([x[j] | j In C]) | C In CD ]);
  Return SR;
EndDefine;   -- StanleyReisnerIdeal

StanleyReisnerIdeal(Delta);

-- specify ring R
Define StanleyReisnerIdeal(R,D)
  z := Indets(R);
  CD := [Diff(1..5,F) | F In D];
  SR := IntersectionList ([Ideal([z[j] | j In C]) | C In CD ]);
  Return SR;
EndDefine;   -- StanleyReisnerIdeal

StanleyReisnerIdeal(CurrentRing, Delta);
----> ideal(x[3]*x[5], x[2]*x[5],
----        x[1]*x[3]*x[4], x[1]*x[2]*x[4])
```

Viceversa, given any squarefree monomial ideal I we can find Δ such that $I_\Delta = I$, again using Alexander duality:

```
I := *** Ideal(x[1]x[2]x[4], x[1]x[3]x[4], x[2]x[4]x[5],
               x[3]x[4]x[5], x[1]x[2]x[5], x[1]x[3]x[5],
               x[2]x[5],     x[3]x[5]) ***;

ADI := AlexanderDual(I);   ADI;    -- or use "MyAlexanderDual"
----> ideal(x[4]*x[5], x[1]*x[5], x[2]*x[3])
W := [Indices(A) | A In Gens(ADI)];  W; -- indices of its gens
----> [[4, 5], [1, 5], [2, 3]]
F := [ Diff(1..5,Ind) | Ind In W ];  F; -- Facets
----> [[1, 2, 3], [2, 3, 4], [1, 4, 5]]
```

Also in this case we write a function:

```
Define Complex(I)
  L := [ Indices(A) | A In Gens(AlexanderDual(I)) ];
  Return [ Diff(1..5,Ind) | Ind In L ];
EndDefine   --- Complex

Complex(I);
----> [[1, 2, 3], [2, 3, 4], [1, 4, 5]]
```

Exercise 1. Check that if the facets of Δ are

$$\mathscr{F}(\Delta) = \{125, 126, 134, 136, 145, 234, 235, 246, 356, 456\}$$

then $I_\Delta = (x_1x_2x_3, x_1x_2x_4, x_1x_3x_5, x_1x_4x_6, x_1x_5x_6, x_2x_3x_6, x_2x_4x_5, x_2x_5x_6, x_3x_4x_5, x_3x_4x_6)$, as said in the example at the end of Sect. 3 in chapter "A Survey on Stanley Depth".

Exercise 2. Compute the Stanley–Reisner ideal of your favourite complex.

Exercise 3. Compute the simplicial complex associated to your favourite square-free ideal.

3 Posets

From now on we denote by P a partially ordered set, poset for short.

Definition 4. A poset ideal I of P is a subset of P which satisfies the following condition: for every $p \in I$ and $q \in P$ with $q \le p$, one has $q \in I$.

Given a finite poset P, one of our goal is to write the list of all the poset ideals. As example we will always (also in the next sections) consider the following poset:

Our representation of this poset is the following:

```
P := [
      [],      -- elements smaller than (1)
      [1],     -- elements smaller than (2)
      [1,2],   -- elements smaller than (3)
      [1,2]    -- elements smaller than (4)
      ];
```

First we define a function to get all the comparable pairs in the poset:

```
Define ComparablePairs(Poset)
  N := Len(Poset);
  Equalities := [ [i,i] | i In 1..N ];
    -- the symbol "><" stands for the cartesian product
  Inequalities := ConcatLists([ Poset[i] >< [i] | i In 1..N ]);
  Return Concat(Equalities, Inequalities);
EndDefine; -- ComparablePairs

ComparablePairs(P);
---->  [[1, 1], [2, 2], [3, 3], [4, 4],
---->   [1, 2], [1, 3], [2, 3], [1, 4], [2, 4]]

-- NEW in CoCoA 5: when I mistype I get a useful error, try
--    ComparableParis(P);
```

Now we use it to write a function that computes the poset ideal "generated" by a list of vertices, i.e. the smallest poset ideal containing the vertices in the list. NB: there is no need to declare types in CoCoA, nor to be consistent ;-)

```
Define PosetIdeal(Poset, X)
  If Type(X) = INT Then
    Return Sorted(Concat(Poset[X], [X]));
  EndIf;
  -- otherwise we expect X to be a LIST:
  U := ConcatLists([ PosetIdeal(Poset, i) | i In X ]);
  Return Sorted(MakeSet(U)); -- "MakeSet" removes duplicates
EndDefine; -- PosetIdeal

-- Examples:
PosetIdeal(P, 2);
----> [1, 2]
PosetIdeal(P, [3,4]);
----> [1, 2, 3, 4]
```

Finally we can use "PosetIdeal" to compute all possible poset ideals in P. Note that this is a "brute force" algorithm trying all subsets, but it is fast enough for our small examples:

```
PosetIdeals := [ PosetIdeal(P, S) | S In subsets(1..Len(P)) ];
-- NEW in CoCoA-5: "indent" prints one element per line
indent(PosetIdeals);

PosetIdeals := MakeSet(PosetIdeals); -- removes duplicates
indent(PosetIdeals);
-- the result is:
[
  [],       -- I_1
  [1, 2, 4],   -- I_2
  [1, 2, 3],   --I_3
  [1, 2, 3, 4],   -- I_4
  [1, 2],   --I_5
  [1]   --I_6
]
```

Exercise 5. Given the posets:

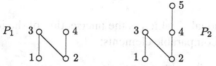

compute the lists of comparable pairs and their poset ideals.

Exercise 6 (not for beginners!!). Write a CoCoA function that, given the "cover relations" of a poset P (the minimal pairs (i, j) with $i \leq j$), computes our representation of P, i.e. the list L with

$$L_i = \{\text{elements smaller than } i\}$$

4 The Hibi Ideal of a Poset

Let $P = \{p_1, \ldots, p_n\}$ be a finite poset.
The ideal $H_P \subseteq R = K[x_1, \ldots, x_n, y_1, \ldots, y_n]$ generated by the monomials

$$u_I = \prod_{p \in I} x_p \prod_{p \notin I} y_q, \quad I \in \mathscr{I}(P)$$

was introduced and studied in [90] and it is called the Hibi ideal of P. For every
$p, q \in P$ set $I_{pq} = (x_p, y_q)$.
    One has:

Theorem 7.

(a) The ideal H_P has a linear resolution.
(b) $H_P = \bigcap_{p \leq q} I_{pq}$.
(c) The multiplicity of R/H_P is the number of the pairs $p, q \in P$ with $p \leq q$.

    Now we consider the example in the previous section, and we calculate the Hibi
Ideal of the Poset P.

```
Use Qxy ::= QQ[x[1..4], y[1..4]];

PosetMonomials :=
  [ Product([x[i] | i In I])
  * Product([y[j] | j In Diff(1..4,I)])
    | I In PosetIdeals ];

H := Ideal(PosetMonomials);    indent(H);
-- the result is
ideal(
  y[1]*y[2]*y[3]*y[4],
  x[1]*x[2]*x[4]*y[3],
  x[1]*x[2]*x[3]*y[4],
  x[1]*x[2]*x[3]*x[4],
  x[1]*x[2]*y[3]*y[4],
  x[1]*y[2]*y[3]*y[4]
)
```

    Consider now the ideal J which is the intersection of the ideals $I_{pq} = (x_p, y_q)$
for every pair p, q of comparable elements:

```
L := ComparablePairs(P);   L;
LL := [Ideal(x[Pair[1]],y[Pair[2]]) | Pair In L];   indent(LL);
J := IntersectionList(LL);   J;

-- From the theory we know that J = H. Let's check it
J = H;

-- the multiplicity =  the number of the comparable pairs
```

```
-- in our example is 9: let's verify it!

Multiplicity(Qxy/H) = Len(ComparablePairs(P));
----> true
```

NB: It has been calculated with CoCoA-4 because it is not yet available in CoCoA-5.0.2.

```
-- Res(Qxy/H);
-- 0 --> Qxy(-6) --> Qxy(-5)^6 --> Qxy(-4)^6 --> R

-- Find the Alexander Dual of the Hibi ideal H
ADH := AlexanderDual(H);      indent(ADH);
```

Since H is the intersection of the ideals (x_p, y_q) with p, q a comparable pair, the Alexander dual of H is the ideal generated by the corresponding products $x_p y_q$. Let's verify it:

```
Ideal([x[CP[1]]*y[CP[2]] | CP In ComparablePairs(P)]) = ADH;
PrimaryDecomposition(H);
```

5 Hibi Rings

Before introducing the Hibi rings we see some procedures to deal with toric rings.

Consider the algebra $\mathbb{Q}[t^3, t^4, t^5] \subseteq \mathbb{Q}[t]$; to find its presentation one has to compute the kernel of the homomorphism

$$\mathbb{Q}[x, y, z] \longrightarrow \mathbb{Q}[t^3, t^4, t^5] \text{ where } x \mapsto t^3, \ y \mapsto t^4, \ z \mapsto t^5$$

In general one has to compute an elimination, but with monomial images we can use the specific function "Toric":

```
L := [3, 4, 5];

Use S ::= QQ[x,y,z, t];  -- general approach to compute Ker
I := Ideal([ Indet(S,i) - t^L[i] | i In 1..3]);
-- that is:  Ideal(x-t^3,  y-t^4,  z-t^5)
Elim([t], I);   -- Ker
----> ideal(y^2 -x*z, -x^2*y +z^2, -x^3 +y*z)

Use QQ[x,y,z];  -- optimised approach for monomial images
RowMat(L);
Toric(RowMat(L));
----> ideal(-y^2 +x*z, x^3 -y*z, -x^2*y +z^2)
```

Another example: compute the kernel of the homomorphism (twisted cubic)

$$\mathbb{Q}[x_1, \ldots x_4] \longrightarrow \mathbb{Q}[s^3, s^2t, st^2, t^3] \subseteq \mathbb{Q}[s, t]$$

$$x_1 \mapsto s^3, \ x_2 \mapsto s^2t, \ x_3 \mapsto st^2, \ x_4 \mapsto t^3$$

```
Use QQ[x[1..4]];
M := Mat([[3,2,1,0],[0,1,2,3]]);  M;
Toric(M);
---->   ideal(x[2]^2 -x[1]*x[3], -x[3]^2 +x[2]*x[4],
----           -x[2]*x[3] +x[1]*x[4])
```

Now come back to Hibi rings.

Let K be a field, $P = \{p_1, \ldots, p_n\}$ a poset, and $\mathscr{I}(P)$ the set of the poset ideals of P. For every $I \in \mathscr{I}(P)$ consider the monomial $u_I = \prod_{p_i \in I} x_i \prod_{p_i \notin I} y_i$.

The Hibi ring of P over K is the toric ring

$$K[P] \subseteq K[x_1, \ldots, x_n, y_1, \ldots, y_n]$$

generated by the monomials $\{u_I : I \in \mathscr{I}(P)\}$ (which are the PosetMonomials). Hibi rings were introduced and studied in [103].

Let $T = K[\{t_I : t_I \in \mathscr{I}(P)\}]$ be the polynomial ring in the variables t_I over K, and

$$\varphi : T \to K[\mathscr{I}(P)]$$

the K-algebra homomorphism with $t_I \mapsto u_I$. Note that $\mathscr{I}(P)$ is a sublattice of the power set of P, and hence it is a distributive lattice.

One has:

Theorem 8.

1. *The ring $K[P]$ is a normal Cohen–Macaulay domain of dimension* $1 + |P|$.
2. *The toric ideal* $\mathrm{Ker}\varphi$ *has a reduced Gröbner basis consisting of the so-called Hibi relations:*

$$t_I t_J - t_{I \cap J} t_{I \cup J} \quad \text{for every pair of incomparable poset ideals } I \text{ and } J.$$

We want to use CoCoA to "verify" Theorem 8 for the poset P introduced in Sect. 3.

We have already computed all the poset ideals, that are:

$$I_1 = \emptyset, \ I_2 = \{1, 2, 4\}, \ I_3 = \{1, 2, 3\}, \ I_4 = \{1, 2, 3, 4\}, \ I_5 = \{1, 2\}, \ I_6 = \{1\}.$$

Thus the only incomparable poset ideals are I_2 and I_3; moreover we have $I_2 \cup I_3 = I_4$ and $I_2 \cap I_3 = I_5$.

We calculate the ideal of the presentation of $K[P]$ using "Toric". A is the matrix whose columns contain the exponent vectors of the monomials.

```
-- recall:
PosetMonomials;

A := Transposed(Mat([Log(M) | M In PosetMonomials]));   A;

Use S ::= QQ[t[1..6]];    -- t[i] --> M[i]
J := Toric(A);    J;         -- Ker
---->  ideal(t[2]*t[3] -t[4]*t[5])
dim(S/J) = Len(P)+1;        -- dim = |P|+1
---->  true
```

Exercise 9. Consider again the poset P_1 and P_2 of Exercise 5.

1. Compute their Hibi ideal and the Hibi ring; play with them by checking dimension, resolution, etc. . . .
2. Do the same with your favourite poset.

6   The Size of a Squarefree Monomial Ideal

For the definition and properties of size and bigsize of an ideal see Sect. 11 of chapter "A Survey on Stanley Depth". We give a procedure to calculate the size of squarefree monomial ideals, by using just the definition of size and the function that determine the primary decomposition of squarefree monomial ideals.

```
Use QQ[x[1..5]];
I := *** Ideal(x[1]x[2], x[2]x[3], x[1]x[4], x[4]x[5]) ***;

-- the ingredients:
PD := PrimaryDecomposition(I); indent(PD);
G := Gens(Sum(PD));   G;
S := Len(Interreduced(G));   S;
indent(Subsets(PD, 2));

-- the function!
Define Size(I)
  PD := PrimaryDecomposition(I);
  S := Len(Interreduced(Gens(Sum(PD))));
  For v := 1 To Len(PD)-1 Do
    Foreach Ideals In Subsets(PD, v) Do -- brute force algorithm
      If Len(Interreduced(Gens(Sum(Ideals)))) = S Then
        Return v + NumIndets(RingOf(I)) - S - 1;
      EndIf;
    EndForeach;
  EndFor;
  Return Len(PD) + NumIndets(RingOf(I)) - S - 1;
EndDefine; -- Size

Size(Ideal(Indets(CurrentRing)));   ---->  0
Size(I);   ---->  1
```

We apply our procedure in the next example:

```
Use QQ[x[1..6]];

I := IntersectionList(
        [ Ideal(x[1],x[2],x[3],x[4]),
          Ideal(x[1],x[6],x[3],x[4]),
          Ideal(x[1],x[5],x[3],x[4]) ]  );
Size(I);  ---->  2
```

Exercise 10.

1. Write a CoCoA function to compute BigSize of a squarefree monomial ideal.
2. Compute Size and BigSize of your favourite squarefree monomial ideal.

7 Stanley Decompositions

Here we show how to compute a Stanley decomposition of R/I, where R is a polynomial ring and I is a monomial ideal [136]. Again this is not optimal, but it is a nice example of CoCoA programming and it works very well for small inputs.

So our base case is given by prime ideals. In this case, when $I = (x_i \mid i \in L)$, the Stanley Decomposition of R/I is simply

$$1 \cdot K[x_i \mid i \notin L]$$

represented as the pair

$$\{(1, \{x_i \mid i \notin L\})\}$$

```
Define IsPrimeMonId(I)
  Foreach g In Gens(I) Do  -- assumes Gens(I) is interreduced
    If Not IsIndet(g) Then Return False; EndIf;
  EndForeach;
  Return True;
EndDefine; -- IsPrimeMonId

-- if I is prime...
Define BaseCase(I)
  Return [ [1, Diff(Indets(RingOf(I)), Gens(I)) ] ];
EndDefine; -- BaseCase
```

The recursive step of the algorithm comes from this short exact sequence

$$0 \to R/(I : p) \xrightarrow{\cdot p} R/I \to R/(I, p) \to 0$$

where we choose a *pivot* p to be an indeterminate properly dividing a generator of I. Here is one possible choice: the first indeterminate of the first reducible generator.

```
Define ChoosePivot(I)
  L := [ g In Gens(I) | Not IsIndet(g) ];
  If L=[] Then Error("I is prime!"); EndIf;
  Return Indet(RingOf(I), First(Indices(L[1])));
EndDefine; -- ChoosePivot
```

Now we have two "simpler" ideals, (I, p) and $I : p$ of which we can recursively compute a Stanley decomposition. Then a Stanley decomposition for I is

$$SD(I) = SD(I, p) \cup p \cdot (SD(I : p)).$$

```
Define InterreducedMonId(I)
  Return Ideal(Interreduced(gens(I)));
EndDefine; -- InterreducedMonId

Define StanleyDecomposition(I)
  If IsPrimeMonId(I) Then
    Return BaseCase(I);
  Else
    p := ChoosePivot(I);
    SD1 := StanleyDecomposition(InterreducedMonId(I+Ideal(p)));
    SD2 := StanleyDecomposition(InterreducedMonId(I:Ideal(p)));
    NewSD2 := [ [p*comp[1], comp[2]] | comp In SD2 ];
    Return concat(SD1, NewSD2);
  EndIf;
EndDefine; -- StanleyDecomposition
```

Now we use our function in the following example:

```
Use QQ[x[1..4]];

indent(StanleyDecomposition(Ideal(x[1]*x[2])));
-- and the result is
[
  [1, [x[2], x[3], x[4]]],
  [x[1], [x[1], x[3], x[4]]]
]
```

Exercise 11.

1. Compute a Stanley decomposition of S/I, where $S = \mathbb{Q}[x_1, x_2, x_3]$ and $I = (x_1^3 x_2, x_1 x_2^3)$; compare it with the one given at the beginning of Sect. 1 of chapter "A Survey on Stanley Depth".
2. Compute a Stanley decomposition of S/I, with I your favourite monomial ideal.

Part II
Edge Ideals

A Beginner's Guide to Edge and Cover Ideals

Adam Van Tuyl

Introduction

Monomial ideals, although intrinsically interesting, play an important role in studying the connections between commutative algebra and combinatorics. Broadly speaking, problems in combinatorics are encoded into monomial ideals, which then allow us to use techniques and methods in commutative algebra to solve the original question. Stanley's proof of the Upper Bound Conjecture [180] for simplicial spheres is seen as one of the early highlights of exploiting this connection between two fields. To bridge these two areas of mathematics, Stanley used square-free monomial ideals.

Over the last decade or so, commutative algebraists have become interested in studying the properties of finite simple graphs through monomial ideals. Fröberg [70], Villarreal [192], and Simis et al. [175] were among the early pioneers in this field. The starting point of these projects is to use the edges of a finite simple graph to construct a monomial ideal, usually called the *edge ideal*, and to study the properties of this monomial ideal using the properties of the graph, and vice versa. In these notes, we provide an introduction to the *edge ideal* and the *cover ideal*, two monomial ideals that can be constructed from a finite simple graph G using its edges, and discuss some current research themes related to these ideals.

Section 1 is devoted to the basics of edge and cover ideals. We start with the definition of these ideals, and work out some of their basic properties. One of the themes in the study of edge and cover ideals is to build a dictionary between graph theory and commutative algebra. To illustrate this dictionary, we explain how

A.V. Tuyl (✉)
Department of Mathematical Sciences, Lakehead University, Thunder Bay, Canada, ON P7B 5E1
e-mail: avantuyl@lakeheadu.ca

A.M. Bigatti et al. (eds.), *Monomial Ideals, Computations and Applications*,
Lecture Notes in Mathematics 2083, DOI 10.1007/978-3-642-38742-5_3,
© Springer-Verlag Berlin Heidelberg 2013

the chromatic number of a graph is encoded algebraically. We also describe how Stanley–Reisner theory can give us information about these ideals.

The goal of Sect. 2 is to introduce the technique of splitting monomial ideals. Splitting a monomial ideal is a technique that originates in a paper of Eliahou and Kervaire [56]. Roughly speaking, a splitting of a monomial ideal allows us to describe its graded Betti numbers in terms of the graded Betti numbers of smaller ideals. Over the last couple of years, this method has proved useful in a number of contexts. We introduce this machinery, and as an application, we prove Fröberg's [70] characterization of edge ideals with a linear resolution.

In Sect. 3, we look at decompositions of powers of the cover ideal. In the first part, we show that some of the associated primes of the powers of the cover ideal correspond to colouring information about the associated graph. In the second part, we turn our attention to the irreducible ideals in the irreducible decomposition of the powers of cover ideals and relate their information to colouring information about the graph. We end with a conjecture about the persistence of associated primes for cover ideals.

At the end of these notes, we provide a brief introduction to the *Macaulay 2* package EdgeIdeals [69], written by C. Francisco, A. Hoefel, and the author. It is hoped that this package will facilitate your own research. We have also include two tutorials that will allow you to start exploring edge and cover ideals using *Macaulay 2*. The tutorials include a number of open problems.

When preparing these notes, I have assumed that the reader is familiar with the basics of Stanley–Reisner ideals, minimal free graded resolutions, and associated primes of ideals. If you need brushing up on some of these topics, I would like to point the reader to [30, 91, 141, 156, 173, 193]. I have not assumed any previous knowledge about graph theory. Instead, I will introduce any terms as needed. However, you may want to have a graph theory textbook handy as you read through these notes.

Given the wealth of research on edge and cover ideals, these notes cannot do adequate justice to this topic. For example, I had no time to look at the interesting problem of determining when a graph is Cohen–Macaulay or sequentially Cohen–Macaulay. My hope is to whet your appetite, and let you explore the field.

If you are interested in learning more, I would recommend that your reading list include Villarreal's textbook [193] (especially Chap. 6), Herzog and Hibi's text [91] (in particular, Chap. 9), the recent survey of Morey and Villarreal [145], and an older survey of Hà and the author [84]. As an aide to help you develop your own projects, scattered throughout the notes are a number of open questions. I would also encourage you to come up with your own questions; one approach is to simply browse a book on graph theory or the latest issue of a graph theory journal and ask yourself if a particular problem or result can be rephrased as an algebraic result. Have fun!

1 The Basics of Edge and Cover Ideals

In this section, we introduce edge and cover ideals. One theme in the study of these ideals is to understand how graph theoretic invariants are encoded algebraically in these ideals. We illustrate this theme by using the chromatic number of a graph as a case study. Because edge and cover ideals are square-free monomial ideals, one can also apply the theory of Stanley–Reisner ideals and simplicial complexes to study these ideals. Using results from Stanley–Reisner theory, we identify other graph theoretic invariants encoded in these ideals. Furthermore, we prove that the edge and cover ideals are dual to each other with respect to Alexander Duality.

1.1 Definitions

Our basic combinatorial object will be a finite simple graph. A *finite graph* G is a pair $G = (V(G), E(G))$ where $V(G) = \{x_1, \ldots, x_n\}$ is the set of *vertices* of G, and $E(G)$ is a collection of two element subsets of $V(G)$, usually called the *edges* of G. A finite graph is simple if we do not allow multiple edges between vertices and we do not allow loops at vertices, i.e., an edge from a vertex x_i to itself.

Example 1. The following pair is a finite simple graph:

$$G = (\{x_1, \ldots, x_5\}, \{\{x_1, x_2\}, \{x_2, x_3\}, \{x_3, x_4\}, \{x_4, x_5\}, \{x_5, x_1\}\}).$$

It is standard practise to represent a graph as a figure. More precisely, we introduce a node for each vertex $x \in V(G)$. We then join two vertices x_i and x_j by a line segment if and only if $\{x_i, x_j\} \in E(G)$. Thus, the above graph G can be represented as the following figure:

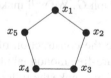

The figure below is not a finite simple graph since it has a loop at the vertex x_1, and three edges from x_4 to x_5:

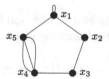

Note that the above graph is sometimes called a *pseudo-graph* or *multi-graph*. We will not consider graphs of this type.

Convention 2. *Throughout these notes, we will assume that G is a finite simple graph, so we will simply call G a graph, and drop the adjectives finite and simple.*

We can study graphs using monomial ideals in a suitable polynomial ring. Suppose we are given a graph $G = (V(G), E(G))$ where $V(G) = \{x_1, \ldots, x_n\}$. We identify the vertices of the graph with the variables in the polynomial ring $R = k[x_1, \ldots, x_n]$. Here, k is some fixed field. (The results of this section are independent of the characteristic of k; however, see Example 44 for a result that depends upon $\mathrm{char}(k)$.)

The graph G is then used to construct two monomial ideals:

Definition 3. Let $G = (V(G), E(G))$ be a graph. The *edge ideal* associated to G is the monomial ideal

$$I(G) = \langle x_i x_j \mid \{x_i, x_j\} \in E(G) \rangle \subseteq R = k[x_1, \ldots, x_n].$$

The *cover ideal* is the monomial ideal

$$J(G) = \bigcap_{\{x_i, x_j\} \in E(G)} \langle x_i, x_j \rangle \subseteq R = k[x_1, \ldots, x_n].$$

Example 4. Let G be as in Example 1. The edge and cover ideals are, respectively:

$$I(G) = \langle x_1 x_2, x_2 x_3, x_3 x_4, x_4 x_5, x_5 x_1 \rangle$$

$$J(G) = \langle x_1, x_2 \rangle \cap \langle x_2, x_3 \rangle \cap \langle x_3, x_4 \rangle \cap \langle x_4, x_5 \rangle \cap \langle x_1, x_5 \rangle$$

$$= \langle x_1 x_3 x_4, x_1 x_3 x_5, x_1 x_2 x_4, x_2 x_4 x_5, x_2 x_3 x_5 \rangle.$$

Remark 5. The term edge ideal originated in a paper of Villarreal [192]. The ideal $J(G)$ is called the cover ideal because its minimal generators correspond to the minimal vertex covers of the graph G. We will make this statement precise later in this section.

Observe that we can reverse the construction of Definition 3 in the following sense. If I is any quadratic square-free monomial ideal in $R = k[x_1, \ldots, x_n]$ of the form $I = \langle x_{1,1} x_{1,2}, \ldots, x_{s,1} x_{s,2} \rangle$ we make the following association:

$$I \mapsto G = (\{x_1, \ldots, x_n\}, \{\{x_{1,1}, x_{1,2}\}, \ldots, \{x_{s,1}, x_{s,2}\}\}).$$

Similarly, if we are given any square-free unmixed height two monomial ideal, i.e., $J = \bigcap_{i=1}^{t} \langle x_{i,1}, x_{i,2} \rangle$, we identify J with the graph:

$$J \mapsto G = (\{x_1, \ldots, x_n\}, \{\{x_{i,1}, x_{i,2}\} \mid i = 1, \ldots, t\}).$$

So, the study of edge and cover ideals hopes to exploit these two one-to-one correspondences:

$$\left\{ \begin{array}{c} \text{EDGE IDEALS} \\ \text{(quadratic} \\ \text{square-free} \\ \text{monomial ideals)} \end{array} \right\} \overset{1-1}{\longleftrightarrow} \left\{ \begin{array}{c} \text{FINITE} \\ \text{SIMPLE} \\ \text{GRAPHS} \end{array} \right\} \overset{1-1}{\longleftrightarrow} \left\{ \begin{array}{c} \text{COVER IDEALS} \\ \text{(height two unmixed} \\ \text{square-free} \\ \text{monomial ideals)} \end{array} \right\}.$$

Remark 6. We have simplified this discussion slightly since one must take into account isolated vertices, or variables of R that do not appear in I or J. We gloss over this technicality here to simplify our discussion.

Given these two one-to-one correspondences, it makes sense to ask:

Question 7. How do the invariants of finite simple graphs relate to the invariants of the edge and cover ideals, and vice versa?

By answering this broad question about the dictionary between two fields, i.e., graph theory and commutative algebra, we may be able to import results from graph theory to help prove algebraic results, and at the same time, export algebraic results to prove graph theory results.

1.2 Colouring Graphs: An Illustration of the Dictionary

To give you a hint of the connection between graph theory and commutative algebra, we will examine the problem of colouring a graph.

Definition 8. A *colouring*[1] of a graph G is an assignment of a colour to each vertex so that adjacent vertices, i.e., vertices joined by an edge receive different colours. The *chromatic number* of a graph G, denoted $\chi(G)$, is the minimum number of colours needed to colour G.

To illustrate this definition, we first introduce two families of graphs that will frequently appear within these notes.

Definition 9. The *clique of size n* with $n \geq 2$, denoted K_n, is the graph with vertex set $V(G) = \{x_1, \ldots, x_n\}$ and edge set $E(G) = \{\{x_i, x_j\} \mid 1 \leq i < j \leq n\}$. We sometimes view an isolated vertex (a vertex with no edges) as a clique of size 1, and denote it by K_1.

The *cycle of size n* with $n \geq 3$, denoted C_n, is the graph with vertex set $V(G) = \{x_1, \ldots, x_n\}$ and edge set $E(G) = \{\{x_1, x_2\}, \{x_2, x_3\}, \{x_3, x_4\}, \ldots, \{x_{n-1}, x_n\}, \{x_n, x_1\}\}$.

Example 10. The graphs below are K_3, K_4, and K_5:

---

[1] I'm using the British–Canadian spelling of colouring, but if you prefer, you can call it a *coloring*. To be consistent, I'll also use *neighbour*.

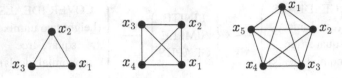

The graphs below are C_3, C_4, and C_5:

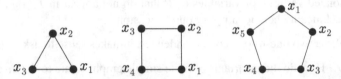

Example 11. The chromatic number of C_5 is three since we can colour it as follows:

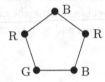

where R represents red, G represent green, and B represents blue.

For both families of graphs, we can easily compute $\chi(G)$:

Lemma 12. *If $G = K_n$, then $\chi(K_n) = n$. If $G = C_n$, then*

$$\chi(C_n) = \begin{cases} 2 \text{ if } n \text{ is even} \\ 3 \text{ if } n \text{ is odd.} \end{cases}$$

Proof. (Exercise)

Application 13. Colouring is a core topic in graph theory; every introductory textbook on graph theory will devote at least one chapter to the topic. It has many practical applications, including the problem of scheduling. To see the connection, suppose we want to schedule a set of exams. Represent each exam by a vertex, and join two vertices if there is a student who must write both exams. For example, suppose we end up with the graph:

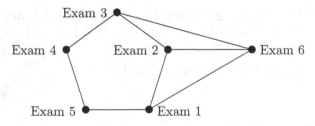

We can colour this graph with three colours (e.g., colour Exam 1 and Exam 3 the same, Exams 5 and 6 are coloured the same, and Exams 2 and 4 are coloured the same). Exams with the same colour can be scheduled at the same time since a student can only take one exam of each colour. The chromatic number of this graph is the smallest number of distinct time slots needed to offer all exams.

Definition 14. A subset $W \subseteq V(G)$ is called an *independent set* if no edge of G has both endpoints in W. A *maximal independent set* is an independent set that is maximal under inclusion.

We make an important observation about a colouring of a graph. Suppose that G has been given a colouring. Then all the vertices of the same colour must form an independent set. This observation leads to the following lemma:

Lemma 15. *Let G be a graph. Then $\chi(G) \leq d$ if and only if $V(G) = C_1 \cup \cdots \cup C_d$ can be partitioned into d independent sets.*

Proof. (Exercise)

There is a concept dual to an independent set:

Definition 16. A subset $W \subseteq V(G)$ is a *vertex cover* if $W \cap e \neq \emptyset$ for all $e \in E(G)$. A vertex cover W is a *minimal vertex cover* if no proper subset of W is a vertex cover.

From the definitions, one can easily prove the following statement:

Lemma 17. *A subset $Y \subseteq V(G)$ is an independent set if and only if $V(G) \setminus Y$ is a vertex cover. In particular, Y is a maximal independent set if and only if $V(G) \setminus Y$ is a minimal vertex cover.*

Proof. (Exercise)

Example 18. Consider the graph C_5 of Example 1. In this graph, $Y = \{x_2, x_5\}$ is an independent set. The complement of Y is the set $V(G) \setminus Y = \{x_1, x_3, x_4\}$. This set is a vertex cover because every edge of C_5 has at least one of its endpoints in $\{x_1, x_3, x_4\}$.

We will now describe how the invariant $\chi(G)$ arises in an algebraic context.

Notation 19. *If $W \subseteq V(G)$, then*

$$x_W := \prod_{x_i \in W} x_i.$$

We first require a lemma that justifies the name cover ideal:

Lemma 20. *Let G be a graph with cover ideal $J(G)$. Then*

$$J(G) = \langle x_W \mid W \subseteq V(G) \text{ is a minimal vertex cover of } G \rangle.$$

Proof. Let $L = \langle x_W \mid W \subseteq V(G)$ is a minimal vertex cover of $G \rangle$.

Let x_W be a generator of L with W a minimal vertex cover. Then, for every edge $e = \{x_i, x_j\} \in E(G)$, we have $W \cap e \neq \emptyset$. So, either $x_i \in W$ or $x_j \in W$. Consequently, either $x_i | x_W$ or $x_j | x_W$, whence $x_W \in \langle x_i, x_j \rangle$. Since e is arbitrary, we have

$$x_W \in \bigcap_{\{x_i, x_j\} \in E(G)} \langle x_i, x_j \rangle = J(G).$$

Conversely, let $m \in J(G)$ be any minimal generator. Note that m must be square-free since $J(G)$ is the intersection of finitely many square-free monomial ideals. So $m = x_{i_1} \cdots x_{i_r}$. Let $W = \{x_{i_1}, \ldots, x_{i_r}\}$. Since $m \in \langle x_i, x_j \rangle$ for each $\{x_i, x_j\} \in E(G)$ either $x_i | m$ or $x_j | m$, and thus, $x_i \in W$ or $x_j \in W$. Thus W is a vertex cover. Let $W' \subseteq W$ be a minimal vertex cover. Because $x_{W'} \in L$ and $x_{W'}$ divides $m = x_W$, we have $m \in L$.

Theorem 21 ([68, Theorem 3.2]). *Let G be a graph with vertices $V(G) = \{x_1, \ldots, x_n\}$. If $J(G)$ is the cover ideal, then*

$$\chi(G) = \min\{d \mid x_{V(G)}^{d-1} = (x_1 \cdots x_n)^{d-1} \in J(G)^d\}.$$

Proof. Let $J = J(G)$.

We first want to show that if $\chi(G) = d$, then $(x_1 \cdots x_n)^{d-1} \in J^d$.

Because $\chi(G) = d$, we have a partition $V_G = C_1 \cup \cdots \cup C_d$ into independent sets (see Lemma 15). Because each C_i is an independent set, the set $W_i = V(G) \setminus C_i$ is vertex cover, and thus $x_{W_i} \in J$. So, $x_{W_1} \cdots x_{W_d} \in J^d$. Each $x_i \in V(G)$ is in exactly one C_i, so each x_i is in exactly $d - 1$ of the W_j. Thus $x_{W_1} \cdots x_{W_d} = (x_1 \cdots x_n)^{d-1} \in J^d$. This shows

$$\chi(G) \geq \min\{d \mid x_{V(G)}^{d-1} = (x_1 \cdots x_n)^{d-1} \in J^d\}.$$

Conversely, let $(x_1 \cdots x_n)^{d-1} \in J^d$. So, we can find d minimal vertex covers $W_1, \ldots, W_d$ such that

$$(x_1 \cdots x_n)^{d-1} = x_{W_1} \cdots x_{W_d} M \in J^d$$

where M is some monomial (possibly $M = 1$). Set

$$C_1 = V(G) \setminus W_1$$
$$C_2 = (V(G) \setminus W_2) \setminus C_1$$
$$C_3 = (V(G) \setminus W_3) \setminus (C_1 \cup C_2)$$
$$\vdots$$
$$C_d = (V(G) \setminus W_d) \setminus (C_1 \cup \cdots \cup C_{d-1}).$$

Then $C_1 \cup \cdots \cup C_d$ is a colouring of G. To see this, first note that each C_i is an independent subset because it is a subset of $V(G) \setminus W_i$. By construction, for each $i \neq j$, we have $C_i \cap C_j = \emptyset$. Finally, for any $x_j \in V(G)$, there is at least one W_i such that $x_j \notin W_i$. Indeed, if $x_j \in W_i$ for all i, then $x_j^d | x_{W_1} \cdots x_{W_d} M = (x_1 \cdots x_n)^{d-1}$, a contradiction. So, $x_j \in C_i = V(G) \setminus W_i$ or $x_j \in C_1 \cup C_2 \cup \cdots \cup C_{i-1}$.

Thus $\chi(G) \leq \min\{d \mid x_{V(G)}^{d-1} = (x_1 \cdots x_n)^{d-1} \in J^d\}$, as desired.

Example 22. Let $G = C_5$. Then a colouring of G is given by

$$V(G) = \underbrace{\{x_1, x_3\}}_{\text{Red}} \cup \underbrace{\{x_2, x_4\}}_{\text{Blue}} \cup \underbrace{\{x_5\}}_{\text{Green}}.$$

Since $\{x_1, x_3\}, \{x_2, x_4\}$, and $\{x_5\}$ are independent sets, the sets $\{x_2, x_4, x_5\}$, $\{x_1, x_3, x_5\}$ and $\{x_1, x_2, x_3, x_4\}$ are vertex covers of G. So $x_2 x_4 x_5$, $x_1 x_3 x_5$, and $x_1 x_2 x_3 x_4 \in J(G)$. It follows that

$$(x_2 x_4 x_5)(x_1 x_3 x_5)(x_1 x_2 x_3 x_4) = (x_1 x_2 x_3 x_4 x_5)^2 \in J(G)^3.$$

Remark 23. Theorem 21 allows us to compute $\chi(G)$ without finding an explicit colouring of G. In the *Macaulay2* [80] package `EdgeIdeals` [69], this formula is used to find the chromatic number. Although we have a simple algorithm for finding the chromatic number, the bottleneck in the computation is finding $J(G)$. As proved in Lemma 20, finding the generators of $J(G)$ is equivalent to finding all the vertex covers of G. However, this is an NP problem!

In graph theory, there is also a notion of a b-fold chromatic number which we can also compute algebraically.

Definition 24. A *b-fold colouring* of a graph G is an assignment to each vertex a set of b distinct colours such that adjacent vertices receive disjoint sets of colours. The *b-fold chromatic number* of G, denoted $\chi_b(G)$, is the minimal number of colours needed in a b-fold colouring of G.

Example 25. The twofold chromatic number of C_5 is five since we can colour it as follows:

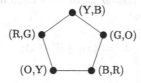

where R represents red, G represent green, B represents blue, O represents orange, and Y represents yellow.

When $b = 1$, then $\chi_b(G) = \chi(G)$. By adapting the proof of Theorem 21, we can derive a much more general result. See the paper [68] for a proof.

Theorem 26. *Let G be a graph with $V(G) = \{x_1, \ldots, x_n\}$. If $J(G)$ is the cover ideal, then*

$$\chi_b(G) = \min\{d \mid x_{V(G)}^{d-b} = (x_1 \cdots x_n)^{d-b} \in J(G)^d\}.$$

Before we turn to the next section, we will end with an open question. The *fractional chromatic number* of a graph G, denoted $\chi_f(G)$, is defined to be

$$\chi_f(G) = \lim_{b \to \infty} \frac{\chi_b(G)}{b}.$$

Then it is known that there exists some integer b such that $\chi_f(G) = \frac{\chi_b(G)}{b}$. Since we can compute $\chi_b(G)$, it is natural to ask:

Question 27. Can we compute $\chi_f(G)$ algebraically?

If you are interested in the fractional chromatic number, see the book of Scheinermann and Ullman [169].[2]

1.3 Associated Simplicial Complexes

Because $I(G)$ and $J(G)$ are square-free monomial ideals, we can use the Stanley–Reisner construction[3] to associate to each ideal a simplicial complex. These simplicial complexes are sometimes useful in our study of the invariants of G using the algebraic properties of $I(G)$ and $J(G)$. We quickly discuss the connection.

Definition 28. A simplicial complex on $V = \{x_1, \ldots, x_n\}$ is a subset Δ of the power set of V such that

(*i*) if $F \in \Delta$, and $G \subseteq F$, then $G \in \Delta$.
(*ii*) $\{x_i\} \in \Delta$ for all i.

One can then associate to Δ a square-free monomial ideal in the ring $R = k[x_1, \ldots, x_n]$:

Definition 29. The *Stanley–Reisner ideal* associated to a simplicial complex Δ is the square-free monomial ideal

$$I_\Delta = \langle x_W \mid W \notin \Delta \rangle.$$

---

[2]Although the book is out-of-print, you can have a free electronic copy if you send the authors a postcard; see http://www.ams.jhu.edu/~ers/fgt/ for details.
[3]For a refresher on Stanley–Reisner rings, see either Stanley [182] or Bruns and Herzog [30].

Notice that this operation can be reversed. That is, if I is a square-free monomial ideal in R, we can associate to I the simplicial complex

$$\Delta(I) = \{W \subseteq V \mid x_W \notin I \text{ and } x_W \text{ square-free}\}.$$

Using some well-known results from Stanley–Reisner theory, we can see how some of the invariants of G are encoded into $I(G)$ and $J(G)$.

Definition 30. Let G be a graph. The *independence complex of G* is the simplicial complex defined by

$$\Delta(G) = \{W \subseteq V(G) \mid W \text{ is an independent set}\}.$$

The following lemma is then a consequence of the definitions:

Lemma 31. *Let G a graph with associated simplicial complex $\Delta(G)$. Then*

$$I(G) = I_{\Delta(G)}.$$

Proof. Let Δ be the simplicial complex associated to $I(G)$. Then

$$
\begin{aligned}
\Delta &= \{W \subseteq V(G) \mid x_W \notin I(G)\} \\
&= \{W \subseteq V(G) \mid \text{for each } \{x_i, x_j\} \in E(G), x_i x_j \nmid x_W\} \\
&= \{W \subseteq V(G) \mid \text{no subset of } W \text{ is an edge}\} = \Delta(G).
\end{aligned}
$$

Definition 32. For a graph G, let

$$\alpha(G) = \max\{|W| \mid W \text{ is a maximal independent set}\}.$$

Recall that elements of a simplicial complex Δ are called *faces*, and that if $F \in \Delta$, then $\dim F = |F| - 1$. Furthermore, we have $\dim \Delta = \max_{F \in \Delta}\{\dim F\}$. Putting together some of the above pieces gives us:

Theorem 33. *For a graph G,*

$$\dim R/I(G) = \alpha(G).$$

Proof. For any simplicial complex Δ, we have $\dim R/I_\Delta = \dim \Delta + 1$ (see, for example, Theorem 1.3 in Chap. 2 of [182]). So,

$$\dim R/I(G) = \dim R/I_{\Delta(G)} = \dim \Delta(G) + 1 = \alpha(G) - 1 + 1 = \alpha(G).$$

For any simplicial complex Δ, the primary decomposition of I_Δ can be described in terms of the *facets* of Δ. Recall that $F \in \Delta$ is a facet if F is a face that is maximal under inclusion. For each facet F, let

$$P_F = \langle x_i \mid x_i \notin F \rangle.$$

Theorem 34. *Let Δ be a simplicial complex with facets $F_1, \ldots, F_t$. Then*

$$I_\Delta = P_{F_1} \cap P_{F_2} \cap \cdots \cap P_{F_t}.$$

Proof. For one proof, see [91, Lemma 1.5.4].

We now apply the above result to the simplicial complex $\Delta(G)$. The facets of $\Delta(G)$ are the maximal independent sets of G. So, if $F \in \Delta(G)$ is a facet,

$$P_F = \langle x_i \mid x_i \notin F \rangle = \langle x_i \mid x_i \in V(G) \setminus F \rangle.$$

But $V(G) \setminus F$ will be a minimal vertex cover of G. Hence, we have

Corollary 35. *Let $W_1, \ldots, W_t$ be the minimal vertex covers of G, and set $\langle W_i \rangle = \langle x_j \mid x_j \in W_i \rangle$. Then*

$$I(G) = \langle W_1 \rangle \cap \cdots \cap \langle W_t \rangle.$$

Example 36. Consider again the graph C_5. Then

$$I(C_5) = \langle x_1, x_3, x_4 \rangle \cap \langle x_1, x_3, x_5 \rangle \cap \langle x_2, x_4, x_5 \rangle \cap \langle x_2, x_3, x_5 \rangle \cap \langle x_1, x_2, x_4 \rangle.$$

As a consequence of Corollary 35, we get an immediate connection between the ideals $I(G)$ and $J(G)$. First, we recall the following important notion:

Definition 37. Let I be a square-free monomial ideal with primary decomposition

$$I = \langle x_{1,1}, x_{1,2}, \ldots, x_{1,s_1} \rangle \cap \langle x_{2,1}, x_{2,2}, \ldots, x_{2,s_2} \rangle \cap \cdots \cap \langle x_{t,1}, x_{1,2}, \ldots, x_{t,s_t} \rangle.$$

The *Alexander Dual* of I, denoted $I^\vee$, is the square-free monomial ideal

$$I^\vee = \langle x_{1,1} x_{1,2} \cdots x_{1,s_1}, x_{2,1} x_{2,2} \cdots x_{2,s_2}, \ldots, x_{t,1} x_{t,2} \cdots x_{t,s_t} \rangle.$$

Corollary 38. *Let G be a graph. Then $I(G)^\vee = J(G)$.*

Example 39. In the previous example, we showed that

$$I(C_5) = \langle x_1, x_3, x_4 \rangle \cap \langle x_1, x_3, x_5 \rangle \cap \langle x_2, x_4, x_5 \rangle \cap \langle x_2, x_3, x_5 \rangle \cap \langle x_1, x_2, x_4 \rangle.$$

So $I(C_5)^\vee = J(G) = \langle x_1 x_3 x_4, x_1 x_3 x_5, x_2 x_4 x_5, x_2 x_3 x_5, x_1 x_2 x_4 \rangle.$

1.4   Additional Constructions

In this section, we have seen two ways to associate to a graph G a monomial ideal. However, these are not the only such constructions. We leave off with two additional constructions and some references.

Construction 40 (Path Ideals). A *path* of length t in a graph G is a set of t distinct vertices $\{x_{i_1}, \ldots, x_{i_t}\}$ such that $\{x_{i_j}, x_{i_{j+1}}\} \in E(G)$ for $j = 1, \ldots, t-1$. The *path ideal* of length of t of G, denoted $I_t(G)$ is the monomial ideal

$$I_t(G) = \langle x_{i_1} \cdots x_{i_t} \mid \{x_{i_1}, \ldots, x_{i_t}\} \text{ is a path of length } t \rangle.$$

Observe that when $t = 2$, a path of length 2 is simply an edge, so $I_2(G)$ is the edge ideal of G. This construction is one way in which we can extend the construction of edge ideals.

   Path ideals were first introduced by Conca and De Negri [40]. Further properties have been developed by He and Van Tuyl [86], Bouchat et al. [27], and Alilooee and Faridi [2].

Construction 41 (Edge and cover ideals of hypergraphs). Our second construction works with a generalization of graphs. Finite simple graphs can be viewed as a special case of a hypergraph. We first recall this definition. Let $\mathscr{X} = \{x_1, \ldots, x_n\}$ be a finite set, and let $\mathscr{E} = \{E_1, \ldots, E_s\}$ be a family of distinct subsets of $\mathscr{X}$. The pair $\mathscr{H} = (\mathscr{X}, \mathscr{E})$ is called a *hypergraph* if $E_i \neq \emptyset$ for each i. The elements of $\mathscr{X}$ are called the *vertices*, while the elements of $\mathscr{E}$ are called the *edges* of $\mathscr{H}$. A hypergraph $\mathscr{H}$ is *simple* if: (1) $\mathscr{H}$ has no loops, i.e., $|E| \geq 2$ for all $E \in \mathscr{E}$, and (2) $\mathscr{H}$ has no multiple edges, i.e., whenever $E_i, E_j \in \mathscr{E}$ and $E_i \subseteq E_j$, then $i = j$. A simple hypergraph is sometimes called a *clutter*. Notice that a hypergraph generalizes the classical notion of a graph; a graph is a hypergraph for which every $E \in \mathscr{E}$ has cardinality two.

   We can then extend the construction of edge and cover ideals to hypergraphs. The *edge ideal* of $\mathscr{H}$ is the ideal

$$I(\mathscr{H}) = \left\langle x_E = \prod_{x \in E} x \;\middle|\; E \in \mathscr{E} \right\rangle \subseteq R = k[x_1, \ldots, x_n].$$

The cover ideal of $\mathscr{H}$ is then defined by:

$$J(\mathscr{H}) = \bigcap_{E \in \mathscr{E}} \langle x \mid x \in E \rangle.$$

   The advantage of this construction is that it gives us a one-to-one correspondence between all square-free monomial ideals and all hypergraphs. (Again, I'm glossing over some details about the isolated vertices.) Properties of these ideals have been studied by Emtander [58], Francisco et al. [68], Hà and Van Tuyl [85] and

Morey et al. [144], among others. These ideals have also been called *facet ideals*; see, for example, Faridi [59].

These constructions are simply the tip of the iceberg. One can make edge rings (see Villarreal [193]) or binomial edge ideals (see Herzog, et al. [92]). Each construction has its own advantage. We end with a very open ended question:

Question 42. Find a new way to associate an algebraic object to a graph. What results can you prove with your correspondence?

2   Splitting Monomial Ideals and Fröberg's Theorem

Let I be a homogeneous ideal in $R = k[x_1, \ldots, x_n]$. Suppose that the graded minimal free resolution[4] of I is given by

$$0 \to \bigoplus_j R(-j)^{\beta_{l,j}(I)} \to \cdots \to \bigoplus_j R(-j)^{\beta_{1,j}(I)} \to \bigoplus_j R(-j)^{\beta_{0,j}(I)} \to I \to 0.$$

Here, $l \leq n$, and $\beta_{i,j}(I)$ is the (i, j)*th graded Betti number of* I. We also let $R(-j)$ denote the polynomial ring shifted in degree j.

An important topic in commutative algebra is the study of graded Betti numbers of an ideal. Since all monomial ideals are homogeneous, it makes sense to ask about the Betti numbers of edge and cover ideals. In fact, approaching this topic, we can hope:

Dream 43. *Describe the graded Betti numbers of* $I(G)$ *and* $J(G)$ *using only the properties of the graph* G.

Unfortunately, we have to call this goal a dream; in reality there exist graphs G where the graded Betti numbers of $I(G)$ depend upon the characteristic of the field k used in the definition of $R = k[x_1, \ldots, x_n]$.

Example 44. Katzman [118] studied how the characteristic of the field affects the graded Betti numbers of $I(G)$. In particular, he showed that the graph G with vertex set $V(G) = \{x_1, x_2, \ldots, x_{11}\}$ and edge set

$$E(G) = \{x_1x_2, x_1x_6, x_1x_7, x_1x_9, x_2x_6, x_2x_8, x_2x_{10}, x_3x_4, x_3x_5, x_3x_7, x_3x_{10},$$

$$x_4x_5, x_4x_6, x_4x_{11}, x_5x_8, x_5x_9, x_6x_{11}, x_7x_9, x_7x_{10}, x_8x_9, x_8x_{10}, x_8x_{11}, x_{10}x_{11}\}$$

has the property that the Betti numbers of $I(G)$ when $\mathrm{char}(k) = 2$ do not agree with the Betti numbers of $I(G)$ when $\mathrm{char}(k) \neq 2$. Moreover, this graph is the smallest such example, where by smallest we mean that no graph on ten or less vertices has

---

[4]If you are not familiar with this notion, see Peeva [156].

this feature. Also see the paper of Dalili and Kummini [46] which also studies how the graded Betti numbers of an edge ideal depend upon the characteristic of the field.

Remark 45. If you are familiar with Hochster's Formula (see, for example, [141, Corollary 5.12]), then it is well known that the graded Betti numbers of a square-free monomial ideal may depend upon the characteristic of the field. Hochster's Formula shows that the graded Betti numbers of an arbitrary monomial ideal can be computed using reduced simplicial homology, which requires information about the field k. However, there are many families of monomial ideals where the Betti numbers are independent of the field; to ask if this fact also applies to edge ideals is a legitimate question that turns out to be false as shown in the previous example.

As an aside, if we take the cover ideal $J(G)$ of the graph G in Example 44, we find an example of a cover ideal whose Betti numbers also depend upon the char(k) [the Betti numbers of $J(G)$ are different if char$(k) = 2$]. To the best of my knowledge, this is the smallest such example. I will leave this as a question:

Question 46. What is the smallest graph G which has the property that the Betti numbers of $J(G)$, the cover ideal, depend upon the characteristic?

Although Dream 43 cannot be realized, we can still ask if some special types of resolutions only depend upon G or if some homological invariants encoded into the resolution only depend upon G. We shall focus on linear resolutions and the regularity of an ideal.

Definition 47. Let I be a homogeneous ideal generated in degree d. Then I has a *linear resolution* if $\beta_{i,i+j}(I) = 0$ for all $j \neq d$.

Another important invariant related to an ideal is the notion of regularity.

Definition 48. The *Castelnuovo–Mumford regularity* (or simply, *regularity*) of an ideal I is

$$\mathrm{reg}(I) = \max\{j - i \mid \beta_{i,j}(I) \neq 0\}.$$

These two notions are linked:

Lemma 49. *Let I be a homogeneous ideal generated in degree d. Then I has a linear resolution if and only if* $\mathrm{reg}(I) = d$.

Proof. (Exercise)

Example 50. Consider the edge ideal of C_5, i.e.,

$$I(C_5) = \langle x_1 x_2, x_2 x_3, x_3 x_4, x_4 x_5, x_5 x_1 \rangle.$$

The graded minimal free resolution of $I(C_5)$ is then given by

$$0 \to R(-5) \to R^5(-3) \to R^5(-2) \to I(C_5) \to 0.$$

So, $\beta_{2,5}(I) = 1$, $\beta_{1,3}(I) = 5$, and $\beta_{0,2}(I) = 5$. Then

$$\text{reg}(I(C_5)) = \max\{2 - 0, 3 - 1, 5 - 2\} = 3.$$

Because $I(C_5)$ is generated in degree 2, and $\text{reg}(I(C_5)) > 2$, this ideal does not have a linear resolution.

Although Dream 43 cannot be answered, perhaps we can answer the following question:

Question 51. Which edge ideals have a linear resolution? Or equivalently, which edge ideals have $\text{reg}(I(G)) = 2$?

This question was first answered by Fröberg [70]. To give the answer, we need some terminology from graph theory.

Definition 52. A *cycle of length* t, denoted $(x_1x_2\cdots x_tx_1)$ is a set of vertices of G such that $\{x_1, x_2\}, \{x_2, x_3\}, \dots, \{x_t, x_1\} \in E(G)$. A *chord* is an edge joining nonadjacent vertices in a cycle. A *minimal cycle* is a cycle with no chords. A *chordal graph* is a graph where all the minimal cycles have length three.

Definition 53. If $G = (V(G), E(G))$ is a graph, the *complementary graph* of G, denoted G^c, is the graph $G^c = (V(G), E(G^c))$ where

$$E(G^c) = \{\{x_i, x_j\} \mid \{x_i, x_j\} \notin E(G)\}.$$

Example 54. Consider the following three graphs:

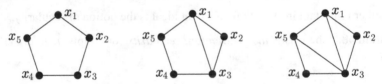

The graph on the left is simply C_5, the cycle of length 5. The second graph has a chord between x_1 and x_3. Note that $(x_1x_2x_3x_4x_5x_1)$ is still a cycle of length 5; however, it is no longer a minimal cycle. The second graph has two minimal cycles, i.e., $(x_1x_2x_3x_1)$ and $(x_1x_3x_4x_5x_1)$. The graph on the right is a chordal graph because all the minimal cycles have length three.

The complements of the above three graphs are, respectively:

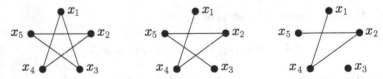

Theorem 55 (Fröberg's Theorem [70]). *Let G be a graph. Then $I(G)$ has a linear resolution if and only if G^c is a chordal graph.*

Example 56. If $G = C_5$, then G^c is not a chordal graph, so $I(G)$ will not have a linear resolution (as we already observed).

Fröberg's Theorem is an important result in the study of edge ideals. There are a number of different proofs of this pivotal result. Besides Fröberg's original paper, you can find different proofs in Dochtermann and Engström [48], Eisenbud et al. [54], Herzog and Hibi [91], and Nevo [150]. We will introduce a technique called *splitting* that enables us to derive some information about the Betti numbers of ideals. As an application, we will show how to use this technique to prove Fröberg's Theorem.

2.1 Splitting Monomial Ideals

Let I be a monomial ideal, and let $\mathscr{G}(I) = \{m_1, \ldots, m_r\}$ denote the set of monomial minimal generators of I. Suppose we partition $\mathscr{G}(I)$ into two sets, i.e.,

$$\mathscr{G}(I) = \{m_1, \ldots, m_s\} \cup \{m_{s+1}, \ldots, m_r\} = \mathscr{G}(J) \cup \mathscr{G}(K).$$

Let $J = \langle m_1, \ldots, m_s \rangle$ and $K = \langle m_{s+1}, \ldots, m_r \rangle$. Note that $I = J + K$.

We then have a short exact sequence:

$$0 \to J \cap K \to J \oplus K \to J + K = I \to 0.$$

If we assume that we are given graded minimal free resolutions of $J \cap K$, J, and K, then we can use the *mapping cone construction*[5] to build a graded resolution of I.

To make this more precise, suppose that we are given the graded minimal free resolutions:

$$0 \to F_{l_1} \to F_{l_1-1} \to \cdots \to F_1 \to F_0 \to J \cap K \to 0,$$

$$0 \to G_{l_2} \to G_{l_2-1} \to \cdots \to G_1 \to G_0 \to J \to 0,$$

and

$$0 \to H_{l_3} \to H_{l_3-1} \to \cdots \to H_1 \to H_0 \to K \to 0.$$

Note that $F_i = \bigoplus_j R^{\beta_{i,j}(J \cap K)}(-j)$, and similarly for G_i and H_i. The mapping cone construction then implies that we can construct the following graded resolution from the above three resolutions:

$$\cdots \to G_2 \oplus H_2 \oplus F_1 \to G_1 \oplus H_1 \oplus F_0 \to G_0 \oplus H_0 \to I \to 0.$$

---

[5]For more details on this construction, see [156, Chap. 1, Sect. 27].

What is important to notice is that this resolution may or may not be a minimal resolution.

Because the Betti numbers of I can only be smaller than the ones given in the above resolution, we always have the following inequality

$$\beta_{i,j}(I) \leq \beta_{i,j}(J) + \beta_{i,j}(K) + \beta_{i-1,j}(J \cap K). \qquad (\star)$$

We are interested in when $(\star)$ is an equality. First let us give a name to this situation.

Definition 57. We call $I = J + K$ a *Betti splitting* if for all $i, j \geq 0$,

$$\beta_{i,j}(I) = \beta_{i,j}(J) + \beta_{i,j}(K) + \beta_{i-1,j}(J \cap K).$$

At first glance, we have no evidence that a Betti splitting should even exist. Let us turn to an example.

Example 58. We return to our favourite example: $G = C_5$. For this case, $I(C_5) = \langle x_1 x_2, x_2 x_3, x_3 x_4, x_4 x_5, x_5 x_1 \rangle$. We partition the generators of $I(G)$ as follows:

$$J = \langle x_1 x_2, x_5 x_1 \rangle \quad \text{and} \quad K = \langle x_2 x_3, x_3 x_4, x_4 x_5 \rangle.$$

Then a simple computation will show that $J \cap K = \langle x_1 x_4 x_5, x_1 x_2 x_3 \rangle$. We now take the minimal graded free resolutions of J, K, and $J \cap K$:

$$0 \to R(-5) \to R^2(-3) \to J \cap K \to 0,$$

$$0 \to R(-3) \to R^2(-2) \to J \to 0,$$

and

$$0 \to R^2(-3) \to R^3(-2) \to K \to 0.$$

The mapping cone construction then gives us a resolution of I:

$$0 \to R(-5) \to R^2(-3) \oplus R^2(-3) \oplus R(-3) \to R^2(-2) \oplus R^3(-2) \to I \to 0.$$

Note that $R^2(-3) \oplus R^2(-3) \oplus R(-3) = R^5(-3)$ and $R^2(-2) \oplus R^3(-2) = R^5(-2)$. So, the mapping cone construction actually gives the minimal free resolution of I, and thus $I = J + K$ is a Betti splitting.

So, Betti splittings exist! Moreover, we get a hint of why they are interesting in the last example. We want to split the ideal I into "smaller" ideals J, K, and $J \cap K$, whose resolutions are either easier to compute, or perhaps we know through induction. We can then build the resolution of the "larger" ideal I.

What we would therefore like are conditions under which a monomial ideal I has a Betti splitting. Splittings of monomial ideals were first introduced by Eliahou and

Kervaire [56]. You are encouraged to look at their criterion for splitting a monomial ideal; we won't reproduce it here because it is somewhat complex and we do not need it. The notion of a splitting was further studied by the C. Francisco, T. Hà, and the author [65]. We will highlight one special case in which monomial ideals can be split.

Theorem 59 (Francisco et al. [65]). *Let $I \subseteq R = k[x_1, \ldots, x_n]$ be a monomial ideal. Fix a variable x_i, and set*

$$J = \langle m \in \mathcal{G}(I) \mid x_i | m \rangle \ \text{ and } \ K = \langle m \in \mathcal{G}(I) \mid x_i \nmid m \rangle.$$

(We call this an x_i-partition of $\mathcal{G}(I)$). If J has a linear resolution, then $I = J + K$ is a Betti splitting.

For any graph G, if $x \in V(G)$, then the *neighbours* of x is the set

$$N(x) = \{y \mid \{x, y\} \in E(G)\}.$$

We can now describe one way to split the edge ideal $I(G)$:

Corollary 60. *Suppose that $G \setminus \{x\}$ (the graph with the vertex x and all adjacent edges removed) is not the graph of isolated vertices. Let $N(x) = \{x_1, \ldots, x_t\}$. Then*

$$I(G) = \langle xx_1, \ldots, xx_t \rangle + I(G \setminus \{x\})$$

is a Betti splitting of $I(G)$.

Proof. We have formed an x-partition of $\mathcal{G}(I(G))$. Use Theorem 59 and the fact that $\langle xx_1, \ldots, xx_t \rangle$ has a linear resolution. $\qquad \blacksquare$

Example 61. Observe that Example 58 is explained by this corollary since $I(C_5) = \langle x_1 x_2, x_1 x_5 \rangle + \langle x_2 x_3, x_3 x_4, x_4 x_5 \rangle$ is an x_1-splitting of $I(C_5)$.

To make use of Corollary 60, we require information about

$$J \cap K = \langle xx_1, \ldots, xx_t \rangle \cap I(G \setminus \{x\}).$$

As we shall see, $J \cap K$ is the sum of a collection of edge ideals constructed from G. We introduce some notation. Fix a vertex $x \in V(G)$, and let $N(x) = \{x_1, \ldots, x_t\}$ be the set of neighbours of x. For $i = 1, \ldots, t$, set

$$G_i = G \setminus (N(x) \cup N(x_i)).$$

We also let $G_{(x)}$ be the graph with edge set

$$\{\{u, v\} \in E(G) \mid \{u, v\} \cap N(x) \neq \emptyset, u \neq x, \text{ and } v \neq x\}.$$

Example 62. We illustrate this example with $G = C_5$. Consider the vertex x_1. So, $N(x_1) = \{x_2, x_5\}$. The graph $G_1 = G \setminus (N(x_1) \cup N(x_2))$ is simply the isolated vertex x_4, while graph $G_2 = G \setminus (N(x_1) \cup N(x_5))$ is the isolated vertex x_3. The graph $G_{(x_1)}$ is the graph:

With this notation, we have

Lemma 63. *Suppose that $G \setminus \{x\}$ is not the graph of isolated vertices. Let $N(x) = \{x_1, \ldots, x_t\}$. Then*

$$\langle xx_1, \ldots, xx_t \rangle \cap I(G \setminus \{x\}) = xI(G_{(x)}) + xx_1 I(G_1) + \cdots + xx_t I(G_t).$$

Proof. See [83].

Example 64. Continuing with the above example, the previous lemma thus implies that if $G = C_5$ and $x = x_1$, then we have

$$\langle x_1 x_2, x_1 x_5 \rangle \cap \langle x_2 x_3, x_3 x_4, x_4 x_5 \rangle = x_1 \langle x_2 x_3, x_4 x_5 \rangle + x_1 x_2 \langle 0 \rangle + x_1 x_5 \langle 0 \rangle$$
$$= \langle x_1 x_2 x_3, x_1 x_4 x_5 \rangle.$$

2.2 Proof of Fröberg's Theorem

We will use the machinery of the last section to prove Fröberg's Theorem (Theorem 55). Interestingly, once we have set up this machinery, the "tough" part of the proof boils down to proving the following graph theoretic result.

Lemma 65. *Suppose G is a graph and $x \in V(G)$ is such that $G \setminus \{x\}$ is not the graph of isolated vertices. Then the following are equivalent:*

(i) G^c *is chordal*
(ii) (a) $(G \setminus \{x\})^c$ *is chordal*
    (b) $G_{(x)}^c$ *is chordal*
    (c) G_i *has no edges.*

Proof. Let $N(x) = \{x_1, \ldots, x_t\}$.

$(i) \Rightarrow (ii)$. Statement (a) comes from the fact that $(G \setminus \{x\})^c = G^c \setminus \{x\}$ is an induced subgraph of G^c, and the chordal property is preserved when passing to induced subgraphs.

We now show that (c) is true. Suppose that there is an i such that $G_i = G \setminus (N(x) \cup N(x_i))$ has an edge, say $\{u, v\}$. By construction, the edges $\{x, u\}, \{x, v\}, \{x_i, u\}$ and $\{x_i, v\}$ do not belong to G. But since $\{x, x_i\}$ and $\{u, v\}$

belong to G, in G^c we will have the minimal four cycle (xux_ivx), contradicting the fact that G^c is chordal. So G_i has no edge.

Finally, we prove (b). The difficulty in proving (b) comes from the fact that $G_{(x)}^c$ is *not* an induced subgraph of G^c. So, suppose $(z_1z_2\cdots z_dz_1)$ is a minimal cycle in $G_{(x)}^c$.

The vertices of $G_{(x)}^c$ are the same as $G_{(x)}$, i.e.,

$$V(G_{(x)}^c) = V(G_{(x)}) = N(x) \cup \{y \mid \{y, x_i\} \in E(G), \ x_i \in N(x) \text{ and } y \notin N(x)\}$$

$$= N(x) \cup N'.$$

We break the proof into four cases depending upon the number of vertices of the cycle $(z_1\cdots z_dz_1)$ that belong to N'.

(A) If $\{z_1,\ldots,z_d\} \subseteq N(x)$, i.e., no vertices belong to N', then the induced graph on G^c on these vertices is still a cycle, so $d = 3$, since G^c is chordal.

(B) If exactly one of $\{z_1,\ldots,z_d\}$ is in N', then again the induced graph on these vertices is still a cycle in G^c, so $d = 3$.

(C) If exactly two of $\{z_1,\ldots,z_d\}$ belong to N', say z_i and z_j, then since z_i and z_i are not adjacent in $G_{(x)}$, $\{z_i,z_j\} \in E(G_{(x)}^c)$. Because this cycle is a minimal cycle, this edge must be part of the cycle. So, after relabelling, we can assume $\{z_1,z_2\}$ is an edge of the cycle, with both vertices in N'. In G^c, the vertex x is adjacent to z_1 and z_2, but none of $\{z_3,\ldots,z_d\}$. So $(z_1xz_2\cdots z_dz_1)$ is a cycle in G^c.

There are now two cases. If $\{z_1,z_2\} \in E(G)$, then $(z_1xz_2\cdots z_dz_1)$ is a minimal cycle of length $d + 1$ in G^c. Since G^c is chordal, $d + 1 = 3$, so $d = 2$. But $(z_1z_2z_1)$ is not a cycle, thus a contradiction. If $\{z_1,z_2\} \notin E(G)$, then $\{z_1,z_2\}$ is a chord in the cycle $(z_1xz_2\cdots z_dz_1)$ in G^c. So $(z_1z_2\cdots z_dz_1)$ is minimal cycle in G^c, whence $d = 3$.

(D) If three or more vertices of the cycle belong to N', then since none of these vertices of N' are adjacent in $G_{(x)}$, these vertices form a clique of size ≥ 3 in $G_{(x)}^c$. A minimal circuit contains a clique only if $d = 3$.

We now show $(ii) \Rightarrow (i)$. By (a), if G^c has a cycle of length ≥ 4, then it must pass through the vertex x. Let $(xz_1z_2\cdots z_dx)$ with $d \geq 3$ be this cycle. We break this into two cases: $d \geq 4$ and $d = 3$.

If $d \geq 4$, then $\{x,z_2\},\ldots,\{x,z_{d-1}\} \notin E(G^c)$ which implies that they all belong to $E(G)$, and thus $\{z_2,\ldots,z_{d-1}\} \subseteq N(x)$. Because the edges $\{x,z_1\}$ and $\{x,z_d\}$ belong to $E(G^c)$, they do not belong to $E(G)$, so $z_1, z_d \notin N(x)$. In addition, the edges $\{z_1,z_{d-1}\}$ and $\{z_d,z_2\}$ are not in $E(G^c)$, so $z_1,z_2 \in N'$, where N' is the same set introduced in the first half of the proof. So $\{z_1,z_d\} \notin E(G_{(x)})$, or equivalently, $\{z_1,z_d\} \in E(G_{(x)}^c)$. Thus, in the graph $G_{(x)}^c$ we have the minimal cycle $(z_1z_2\ldots z_dz_1)$ of length at least four, which contradicts (b).

Now suppose $d = 3$, i.e., suppose $(xz_1z_2z_3x)$ is a minimal four cycle in G^c. So $\{x,z_2\}$ and $\{z_1,z_3\}$ must be edges in $E(G)$ since they are not edges of G^c. Note that $z_1, z_3 \notin N(x)$. Because $z_2 \in N(x)$ and because $G \setminus (N(x) \cup N(z_2))$ has no edges, either $z_1 \in N(z_2)$ or $z_3 \in N(z_2)$. Indeed, if $z_1, z_3 \notin N(z_2)$, then $\{z_1,z_3\}$ would be an edge of $G \setminus (N(x) \cup N(z_2))$. So, either $\{z_1, z_2\}$ or $\{z_2, z_3\}$ are edges of G. But this contradicts the fact that both edges $\{z_1, z_2\}$ and $\{z_2, z_3\}$ belong to G^c. So G^c has no minimal cycle of length ≥ 4, and thus, G^c must be chordal.

We now come to the main result of the section:

Proof. (of Fröberg's Theorem [Theorem 55]) We do induction on $|V(G)|$. If $|V(G)| \leq 3$, one simply checks all possible graphs. If G is the graph

$$G = (\{x, x_1, \ldots, x_t\}, \{\{x, x_1\}, \{x, x_2\}, \ldots, \{x, x_t\}\})$$

for some $t \geq 1$, then $I(G) = x(x_1, \ldots, x_t)$ has a linear resolution and G^c is chordal.

So, we can assume that $|V(G)| \geq 4$ and that there is a vertex $x \in V(G)$ such that $G \setminus \{x\}$ is not the graph of isolated vertices. Thus, there is a Betti splitting of $I(G)$, whence

$$\beta_{i,j}(I(G)) = \beta_{i,j}((xx_1, \ldots, xx_t)) + \beta_{i,j}(I(G \setminus \{x\})) + \beta_{i-1,j}(L) \quad (\star)$$

where $L = xI(G_{(x)}) + xx_1I(G_1) + \cdots + xx_tI(G_t)$.

Let us suppose that $I(G)$ has a linear resolution. This implies that $(xx_1, \ldots, xx_t)$, $I(G \setminus \{x\})$, and L all have a linear resolution. By induction $(G \setminus \{x\})^c = G^c \setminus \{x\}$ is chordal. Because L has a linear resolution, we must have $L = xI(G_{(x)})$ since L cannot have generators in degree three and four. So, that means G_i has no edges for $i = 1, \ldots, t$. Finally, since $L = xI(G_{(x)})$ has a linear resolution, then so must $I(G_{(x)})$, and thus by induction $G_{(x)}^c$ is chordal. Now apply Lemma 65 to deduce that G^c is chordal.

Conversely, suppose G^c is chordal. The ideal $(xx_1, \ldots, xx_t)$ always has a linear resolution. By Lemma 65, $(G \setminus \{x\})^c$ is chordal, thus by induction, $I(G \setminus \{x\})$ has a linear resolution. Again, Lemma 65 implies that $L = xI(G_{(x)})$ with $G_{(x)}^c$ chordal. So, by induction, L has linear resolution. Hence, our splitting formula $(\star)$ implies $I(G)$ has a linear resolution.

Remark 66. Since the appearance of Fröberg's Theorem, there have been a number of interesting generalizations. Eisenbud et al. [54] describe how long the resolution will stay linear (sometimes called the $N_{2,p}$ property) in terms of the length of the smallest induced cycle in the complement of G [also see Tutorial Exercises 19 and 20 in the chapter "Edge Ideals Using Macaulay2" for more details]. Some additional results in this direction were discovered by Fernández-Ramos and Gimenez [61]. Finding a hypergraph version of Fröberg's Theorem appears much more difficult, in part, because the characteristic of the field becomes relevant. See [57, 85, 198] for some known facts about this case.

2.3   Additional Comments and Open Questions

2.3.1   Splitting Monomial Ideals

We have only scratched the surface when it comes to splitting ideals. As mentioned, this idea first arose in a paper of Eliahou and Kervaire [56]. Further properties were developed by Francisco et al. [65].

You should see splitting as a useful tool to deduce results about monomial ideals. For example, T. Hà and myself found a number of results about the edge ideals of graphs and hypergraphs [83, 85]. Francisco [64], Fatabbi [60] and Valla [189] have all used the splitting of monomial ideals to study the ideal of fat points in projective space $\mathbb{P}^n$. Hopefully this section has shown you that the technique of splitting a monomial ideal should belong to your toolbox.

There are many questions that can still be asked. In this section, we looked at one way to split $I(G)$. It makes sense to ask:

Question 67. Can we find other splittings of the edge ideal $I(G)$?

As an example of another answer to this question, T. Hà and myself [83] introduced what we called *edge splittings*, i.e., finding conditions under which

$$I(G) = \langle x_i x_j \rangle + I(G \setminus e)$$

is a splitting; here $G \setminus e$ is the graph G with the edge $e = \{x_i, x_j\}$ removed. This splitting lead to a nice recursive formula for the graded Betti of chordal graphs [85]. Hibi, Kimura, and Murai where then able to exploit this formula in their study of f-vectors (see [105]); this formula was also used by Kimura [121]. There may be other ways to split $I(G)$.

As you may have noticed, we have not said much about the cover ideal $J(G)$ in this section. Of course, we can ask

Question 68. Can we find splittings of the cover ideal $J(G)$?

I know of only one special case where a splitting of $J(G)$ has been found. C. Francisco, T. Hà, and myself [65] showed that $J(G)$ can be split if $R/I(G)$ is Cohen–Macaulay, and the graph G is bipartite. It would be nice to find other natural ways to split $J(G)$.

2.3.2   Regularity of Edge Ideals

The regularity of ideals is a fascinating topic. Even in the case of edge ideals, there is much more we would like to know.

As noted, Fröberg's Theorem classifies all the edge ideals which have regularity two. Herzog, Hibi, and Zheng have given a very interesting generalization of this theorem:

Theorem 69 ([93]). *If G^c is chordal, then $I(G)^s$ has a linear resolution for all $s \geq 1$. In particular, $\mathrm{reg}(I(G)^s) = 2s$.*

Nevo and Peeva [151] have made a conjecture about the regularity of edge ideals, which generalizes this theorem.

Conjecture 70. Suppose that G^c has no minimal four-cycles. Then there exists an integer t such that $I(G)^s$ has a linear resolution for all $s \geq t$. In particular, $\mathrm{reg}(I(G)^s) = 2s$ for $s \gg 0$.

Nevo [150] and Hoefel and Whieldon [106] have given some partial evidence for this conjecture. Nevo has shown that if the graph G has the property that G^c has no minimal four cycle, and G has no claw, then $I(G)^2$ has a linear resolution. A *claw* is any induced subgraph of the form

Hoefel and Whieldon have shown that if $G = C_n^c$, the complement of an n-cycle, then $I(G)^2$ has a linear resolution, even though $I(G)$ does not.

Finally, if you are interested in learning more about the regularity of edge ideals, the following papers should be part of your reading list: [44, 85, 125, 143, 145, 190, 196, 197, 203].

2.3.3 Simplicial and Cellular Resolutions

We have focused primarily on the case of computing graded Betti numbers (or bounding these invariants, as in the case of regularity). There has also been some interest in describing the structure (e.g., describing the maps) of the minimal free resolutions of edge and cover ideals.

In addition, one would like to determine if there exists a simplicial complex or cell complex that supports a particular resolution. It is beyond the scope of these notes to define these terms, so we point the reader to the textbook of Peeva [156]. We will simply say that one wishes to identify topological objects that encode the structure of the minimal free resolution of a monomial ideal.

To date, most investigations have focused on the structure of the minimal free resolutions of edge ideals. Biermann [20] looked at the resolutions of $I(C_n^c)$, i.e., the edge ideal of the complement of cycles; Chen [34] and Horwitz [107] examined the case of ideals with linear resolutions, i.e., ideals of the form $I(G)$ with G^c chordal; Corso and Nagel [43, 44] considered the case of Ferrers graph; and Dochtermann and Engström [49] studied the cellular resolutions of co-interval graphs. I do not know of any work on the structure of cover ideals of graphs.

3 Colouring Graphs and Decomposing Cover Ideals

Let $I \subseteq R = k[x_1, \ldots, x_n]$ be any ideal. Recall that a prime ideal P is *associated* to an ideal I if there exists an element $m \in R$ such that $I : \langle m \rangle = P$. The *set of associated primes* is then the set

$$\mathrm{Ass}(I) = \{P \mid P \text{ is associated to } I\}.$$

In the case that I is a monomial ideal, all the associated prime ideals must also be monomial. Furthermore, the only prime monomial ideals are those of the form $\langle x_{i_1}, \ldots, x_{i_r} \rangle$.

In the case that $I = I(G)$ or $I = J(G)$, we already know the associated primes from Sect. 1:

$$\mathrm{Ass}(I(G)) = \{\langle x_{i_1}, \ldots, x_{i_t} \rangle \mid \{x_{i_1}, \ldots, x_{i_r}\} \text{ is a minimal vertex cover}\}$$
$$\mathrm{Ass}(J(G)) = \{\langle x_i, x_j \rangle \mid \{x_i, x_j\} \in E(G)\}.$$

Note that the variables that generate the associated primes of $I(G)$ and $J(G)$ are telling us something about the graph G at the corresponding vertices.

Since we already know $\mathrm{Ass}(I)$ for $I = I(G)$ and $I = J(G)$, we can ask:

Problem 71. Suppose $I = I(G)$ or $I = J(G)$. Describe the sets $\mathrm{Ass}(I^s)$ as s varies.

We will focus on the case that $I = J(G)$. For the case of the edge ideal, see the papers of Chen et al. [33] and Martinez-Bernal et al. [138].

As a teaser, we already saw in Sect. 1 that the chromatic number of G was encoded into information about the powers of $J(G)$. We will discover that additional colouring information will be encoded into the powers of $J(G)$.

Before proceeding, we first point out that we will be abusing notation slightly. We shall let P denote both the monomial ideal $\langle x_{i_1}, \ldots, x_{i_t} \rangle$ and the subset $\{x_{i_1}, \ldots, x_{i_t}\} \subseteq V(G)$. It will be clear from the context whether P is an ideal or a subset of $V(G)$.

In addition, we need the notion of an *induced subgraph*. Let $P \subseteq V(G)$. The induced graph of G on P is the graph

$$G_P = (P, E(G_P)) = (P, \{\{x_i, x_j\} \in E(G) \mid \{x_i, x_j\} \subseteq P\}).$$

Example 72. Consider the graph $G = C_5$, and let $P = \{x_2, x_3, x_5\}$. Then the induced graph G_P is the graph:

3.1   Powers of Cover Ideals: Associated Primes

We begin with a lemma that reduces our problem to determining if the maximal ideal $\langle x_1, \ldots, x_n \rangle$ is an associated prime of $J(G)^s$.

Lemma 73. *The following are equivalent:*

(i) $P = \langle x_{i_1}, \ldots, x_{i_r} \rangle \in \mathrm{Ass}(J(G)^s)$ *with* $J(G) \subseteq k[x_1, \ldots, x_n]$.
(ii) $P = \langle x_{i_1}, \ldots, x_{i_r} \rangle \in \mathrm{Ass}(J(G_P)^s)$ *with* $J(G_P) \subseteq k[x_{i_1}, \ldots, x_{i_r}]$.

Proof. The details are worked out in [68] using the properties of localization.

We take a detour to introduce some more graph theory:

Definition 74. A graph G is *critically s-chromatic* if $\chi(G) = s$, and for every $x \in V(G)$, $\chi(G \setminus \{x\}) < s$.

Example 75. Let $G = C_n$ be the n-cycle with n odd. Then G is a critically 3-chromatic graph since $\chi(G) = 3$, but if we remove any vertex x, then $\chi(G \setminus \{x\}) = 2$.

Example 76. Let $G = K_n$ be the clique of size n. Then G is a critically n-chromatic graph since $\chi(G) = n$, but if we remove any vertex x, then $G \setminus \{x\} = K_{n-1}$, and thus $\chi(G \setminus \{x\}) = n - 1$.

Remark 77. You should be able to convince yourself that the only critically 1-chromatic graph is the graph of an isolated vertex, and the only critically 2-chromatic graph is K_2. The only critically 3-chromatic graphs are precisely the graphs $G = C_n$ with n odd. However, for $s \geq 4$, there is no known classification of critically s-chromatic graphs.

As the next theorem shows, some of the associated primes of $J(G)^s$ are actually detecting induced subgraphs that are critically $(s + 1)$-chromatic.

Theorem 78. *Let G be a graph and suppose $P \subseteq V(G)$ is such that G_P is critically $(s + 1)$-chromatic. Then*

(1) $P \notin \mathrm{Ass}(J(G)^d)$ *for* $1 \leq d < s$.
(2) $P \in \mathrm{Ass}(J(G)^s)$.

Proof. By Lemma 73, we can assume that $G = G_P$.

(1) Suppose that $P \in \mathrm{Ass}(J(G)^d)$ for some $d < s$. Thus, there exists some monomial $m \notin J(G)^d$ such that $J(G)^d : \langle m \rangle = P$.

We first note that $m \mid (x_1 \cdots x_n)^{d-1}$. If not, then there is some x_i such that $x_i^d \mid m$. Because $x_i m \in J(G)^d$, we can find d vertex covers $W_1, \ldots, W_d$ such that $x_i m = x_{W_1} \cdots x_{W_d} M \in J(G)^d$ for some monomial M. Since x_i appears at least $d + 1$ times on the left, it must appear the same number of times on the right. Because each x_{W_j} is square-free, this means that $x_i \mid M$. We then have $m = x_{W_1} \cdots x_{W_d} (M/x_i) \in J(G)^d$, contradicting the fact that $m \notin J(G)^d$.

Because $J(G)^d : \langle m \rangle = P$, we have $x_1 m \in J(G)^d$. Note that $x_W = x_2 x_3 \cdots x_n \in J(G)$ since $W = \{x_2, \ldots, x_n\}$ is a vertex cover. So, $x_1 m x_W \in J(G)^{d+1}$, and moreover, $x_1 m x_W$ will divide $(x_1 \cdots x_n)^d$. So, $(x_1 \cdots x_n)^d \in J(G)^{d+1}$. By Theorem 21, this means that $\chi(G) \leq d + 1 < s + 1$. But this contradicts the fact that $\chi(G) = (s+1)$.

We now prove (2). We are given

$$\chi(G) = \min\{t \mid (x_1 \cdots x_n)^{t-1} \in J(G)^t\} = s + 1$$

so $m = (x_1 \cdots x_n)^{s-1} \notin J(G)^s$. In other words, we have, $J(G)^s : \langle m \rangle \subsetneq \langle 1 \rangle$, and hence $J(G)^s : \langle m \rangle \subseteq \langle x_1, \ldots, x_n \rangle$. We will now show that $J(G)^s : \langle m \rangle \supseteq \langle x_1, \ldots, x_n \rangle$; the conclusion will then follow from this fact.

Since G is critically $(s+1)$-chromatic, for each $x_i \in V(G)$, $\chi(G \setminus \{x_i\}) = s$. Let

$$V(G \setminus \{x_i\}) = C_1 \cup \cdots \cup C_s$$

be the s colouring of $V(G \setminus \{x_i\})$. Then

$$V(G) = C_1 \cup \cdots \cup C_s \cup \{x_i\}$$

is an $(s+1)$-colouring of G.

For $j = 1, \ldots, s$, set

$$W_j = C_1 \cup \cdots \cup \widehat{C_j} \cup \cdots \cup C_s \cup \{x_i\}.$$

Each W_j is a vertex cover, so $x_{W_j} \in J(G)$. Thus

$$\prod_{j=1}^{s} x_{W_j} \in J(G)^s.$$

But $\prod_{j=1}^{s} x_{W_j} = (x_1 \cdots x_n)^{s-1} x_i$. Thus, $x_i \in J(G)^s : \langle m \rangle$. This is true for each $x_i \in V(G)$, whence $\langle x_1, \ldots, x_n \rangle \subseteq J(G)^s : \langle m \rangle \subseteq \langle x_1, \ldots, x_n \rangle$, as desired.

Remark 79. It is believed (see Sect. 3.3 for more details) that Theorem 78 (2) can be strengthened to (2′) $P \in \mathrm{Ass}(J(G)^d)$ for all $d \geq s$.

Example 80. We consider the following graph

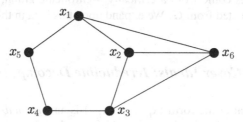

Note that the induced graph on $\{x_1, x_2, x_6\}$ is a K_3 (and C_3), a critically 3-chromatic graph. So $P = \langle x_1, x_2, x_6 \rangle$ is in $\mathrm{Ass}(J(G)^2)$, but not in $\mathrm{Ass}(J(G))$. Similarly, since the induced graph on $\{x_1, x_2, x_3, x_4, x_5\}$ is a C_5, we will have $\langle x_1, x_2, x_3, x_4, x_5 \rangle \in \mathrm{Ass}(J(G)^2)$.

When $s = 2$, we can find a converse of Theorem 78. In fact, we can give a complete characterization of the associated primes of $J(G)^2$; this result first appeared in [67].

Theorem 81. *Let G be a graph.*
A prime ideal $P = \langle x_{i_1}, \ldots, x_{i_t} \rangle \in \mathrm{Ass}(J(G)^2)$, if and only if:

1. $P = \langle x_{i_1}, x_{i_2} \rangle$, and $\{x_{i_1}, x_{i_2}\} \in E(G)$, or
2. t is odd, and the induced graph on $\{x_{i_1}, x_{i_2}, \ldots, x_{i_t}\}$ is an induced cycle of G.

Example 82. By Theorem 81, we can explicitly write out all the elements of $\mathrm{Ass}(J(G)^2)$ for the graph G of Example 80:

$$\mathrm{Ass}(J(G)^2) = \{\langle x_1, x_2 \rangle, \langle x_2, x_3 \rangle, \cdots, \langle x_3, x_6 \rangle, \langle x_1, x_2, x_6 \rangle, \langle x_2, x_3, x_6 \rangle,$$
$$\langle x_1, x_2, x_3, x_4, x_5 \rangle, \langle x_1, x_6, x_3, x_4, x_5 \rangle\}$$

Remark 83. By Theorem 81, the associated primes of $J(G)^2$ are related to the odd induced cycles in the graph. This gives a method to identify all the odd induced cycles in a graph; in fact, this is the procedure used in the *Macaulay 2* `EdgeIdeals` package. Since odd induced graphs play an important role in the classification of perfect graphs (see [35]), we can exploit the associated primes of $J(G)^2$ to determine if a graph is perfect. Again, see [67] for all the details.

Unfortunately, the converse of Theorem 78 is false in general; that is, if $P \in \mathrm{Ass}(J(G)^s)$, but $P \notin \mathrm{Ass}(J(G)^d)$ with $1 \le d < s$, then the graph G_P is not necessarily a critically $(s + 1)$-chromatic graph.

Example 84. If we consider the graph of Example 80, then the prime ideal $P = \langle x_1, x_2, x_3, x_4, x_5, x_6 \rangle \in \mathrm{Ass}(J(G)^3)$ but not in $\mathrm{Ass}(J(G))$ or $\mathrm{Ass}(J(G)^2)$. However, the graph $G = G_P$ is not critically 4-chromatic. In fact, $\chi(G) = 3$.

What is happening here is that the colouring information in the associated primes is too "crude". We need to decompose the ideal $J(G)^s$ differently to extract the colouring information. The ideal $P = \langle x_1, x_2, x_3, x_4, x_5, x_6 \rangle \in \mathrm{Ass}(J(G)^3)$ in the above example does come from a critically 4-chromatic graph, but it "lives" in a larger graph constructed from G. We expand upon this idea in the next section.

3.2 Powers of Cover Ideals: Irreducible Decomposition

Any monomial ideal of the form $\langle x_{i_1}^{a_{i_1}}, \ldots, x_{i_t}^{a_{i_t}} \rangle$ is an *irreducible monomial ideal*. A monomial ideal can then be decomposed into irreducible monomial ideals:

Theorem 85. *Every monomial ideal I has a unique irredundant decomposition into irreducible ideals; i.e., we can write I uniquely as*

$$I = m_1 \cap \cdots \cap m_t$$

where each m_i is an irreducible monomial ideal.

Proof. See [141, Theorem 5.27]. $\qquad\blacksquare$

The next lemma is the basis of an algorithm to find this irreducible decomposition.

Lemma 86. *Let I be a monomial ideal. If m is a minimal generator of I and $m = m_1 m_2$ with $\gcd(m_1, m_2) = 1$, then*

$$I = (I + \langle m_1 \rangle) \cap (I + \langle m_2 \rangle).$$

Example 87. If $I = \langle x^2, xy, y^2 \rangle$, we can decompose it as

$$I = \langle x^2, xy, y^2, x \rangle \cap \langle x^2, xy, y^2, y \rangle = \langle x, y^2 \rangle \cap \langle x^2, y \rangle.$$

This example gives a hint of why an irreducible decomposition will be more useful. Note that if $J = \langle x, y \rangle$, then $I = J^2$. Then not only can we read off the associated primes, e.g., $\mathrm{Ass}(J^2) = \{\langle x, y \rangle\}$, but the irreducible decomposition is another way to express the original ideal.

We now take a detour to introduce some more graph theory:

Definition 88. Given a graph $G = (V(G), E(G))$ and integer $s \geq 1$, the *s-th expansion of G*, denoted G^s, is the graph constructed from G as follows: (a) replace each $x_i \in V(G)$ with a clique of size s on the vertices $\{x_{i,1}, \ldots, x_{i,s}\}$, and (b) two vertices $x_{i,a}$ and $x_{j,b}$ are adjacent in G^s if and only if x_i and x_j were adjacent in G.

Example 89. We illustrate this example when $G = C_4$, and we construct G^2. Recall that C_4 is the graph:

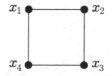

Then the second expansion of G is the graph:

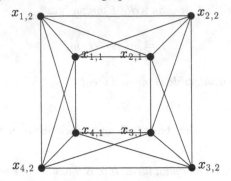

We now come to our main result:

Theorem 90. *Let G be a graph with cover ideal $J(G)$.*
Then $\langle x_{i_1}^{a_{i_1}}, \ldots, x_{i_r}^{a_{i_r}} \rangle$ appears in the irreducible decomposition of $J(G)^s$ if and only if the induced graph on

$$\{x_{i_1,1}, \ldots, x_{i_1,s-a_{i_1}+1}, \ldots, x_{i_r,1}, \ldots, x_{i_r,s-a_{i_r}+1}\}$$

in G^s is a critically $(s+1)$-chromatic graph.

The proof is a mixture of a number of ingredients. It relies on generalized Alexander duality, polarization and depolarization of monomial ideals, and a result of Sturmfels and Sullivant [185]. We have only stated it for edge ideals of graphs, but it works also for edge ideals of hypergraphs, i.e., any square-free monomial ideal.

Example 91. Let us return to Example 80 and explain why the prime ideal $\langle x_1, x_2, x_3, x_4, x_5, x_6 \rangle$ appears in $\mathrm{Ass}(J(G)^3)$. If we look at the irreducible decomposition of $J(G)^3$ (which we can compute using *Macaulay 2*), one of the irreducible monomial ideals that appears is the ideal

$$\langle x_1^3, x_2^3, x_3^3, x_4^3, x_5^2, x_6^3 \rangle.$$

So, we need to look at the induced graph on

$$\{x_{1,1}, x_{2,1}, x_{3,1}, x_{4,1}, x_{5,1}, x_{5,2}, x_{6,1}\}$$

in G^3. This graph looks like:

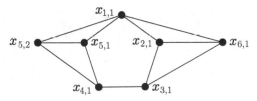

You can now convince yourself that this graph is critically 4-chromatic.

3.3 Persistence of Primes and a Conjecture

We end with a conjecture about the persistence of primes.

Definition 92. An ideal I in a Noetherian ring R has the *persistence property* if

$$\text{Ass}(I^s) \subseteq \text{Ass}(I^{s+1}) \text{ for all } s \geq 1.$$

Not every ideal has the persistence property, and in fact, proving that an ideal has this property can be quite difficult. With respect to edge and cover ideals, we have the following results:

Theorem 93. (a) (See [138]) for any graph G, the edge ideal $I(G)$ has the persistence property.
(b) (See [68]) if G is a chordal graph,[6] then $J(G)$ has the persistence property.

What is interesting about the proof of (a) found in [138] is that one needs a classical result from graph theory due to Berge on matchings in a graph to prove the persistence property. The paper [138] is a good example of using results from graph theory to prove an interesting algebraic result.

Computer experiments have suggested that Theorem 93(b) will hold for any graph G, i.e., the cover ideal $J(G)$ for any G also has the persistence property. Unlike the case of edge ideals, we need a graph theory result that appears unknown. We state the "missing" graph theory result.

Definition 94. Let $W \subseteq V(G)$. The *expansion* of G at W, denoted $G[W]$ is the graph obtained by replacing $x_i \in W$ with the edge $\{x_{i,1}, x_{i,2}\}$ and joining to these two vertices all the vertices to which x_i was joined.

The following conjecture is found in [66]:

Conjecture 95. Suppose that G is a critically s-chromatic graph. Then there exists a $W \subseteq V(G)$ such that $G[W]$ is a critically $(s + 1)$-chromatic graph.

Example 96. The conjecture is true for all odd cycles (i.e., the critically 3-chromatic graphs; see [66] for a proof) and for graphs whose fractional chromatic number $\chi_f(G)$ (see Sect. 1) is "close" to $\chi(G)$, i.e., $\chi(G) - 1 < \chi_f(G) \leq \chi(G)$. We illustrate the conjecture with the graph $G = C_5$. If we expand this graph at the vertices $W = \{x_2, x_5\}$, we get the graph

---

[6]The result holds for a larger class of graphs called *perfect graphs*.

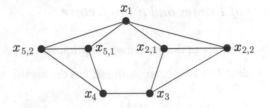

This graph is critically 4-chromatic.

Francisco, Hà, and myself then proved that if this graph theory conjecture is true, then one could prove that every cover ideal has the persistence property.

Theorem 97. *If Conjecture 95 is true, then every cover ideal $J(G)$ has the persistence property.*

Since the conjecture is true for $s = 2$ and $s = 3$, we know that for all graphs G, we have the following containments:

$$\text{Ass}(J(G)) \subseteq \text{Ass}(J(G)^2) \subseteq \text{Ass}(J(G)^3).$$

We end with an open ended question. For any hypergraph, one can formulate a hypergraph version of Conjecture 95 and Theorem 90. Since the cover ideals of hypergraphs are in one-to-one correspondence with all square-free monomial ideals, do we really have:

Question 98. Is it true that all square-free monomial ideals have the persistence property?

Note added in proof: In January 2013, Kaiser, Stehlik and Skrekovski posted a preprint [117] that contains a counterexample to Conjecture 95. The graph that provides a counterexample can also be used to give a negative answer to Question 98.

Acknowledgements I would like to thank the organizers of MONICA, Anna M. Bigatti, Philippe Gimenez, and Eduardo Sáenz-de-Cabezón, for the invitation to participate in this conference. As well, I would like to thank all the participants for stimulating discussions and their feedback. I would also like to thank Ben Babcock, Ashwini Bhat, Jen Biermann, Chris Francisco, Tai Hà, Andrew Hoefel, and Ştefan Tohăneanu for their feedback on preliminary drafts. The author was supported in part by an NSERC Discovery Grant.

Edge Ideals Using Macaulay2

Adam Van Tuyl

1 The *Macaulay 2* Package `EdgeIdeals`

Computer algebra systems, like *Macaulay 2* [80], Singular [47], and CoCoA [39], have become essential tools for many mathematicians in commutative algebra and algebraic geometry. These systems provide a "laboratory" in which we can experiment and play with new ideas. From these experiments, a researcher can formulate new conjectures, and hopefully, new theorems. Computer algebra systems are especially good at dealing with monomial ideals. As a consequence, the study of edge and cover ideals is well suited to experiments using computer algebra systems.

The purpose of this section is to familiarize the user with the package `EdgeIdeals` that was written by C. Francisco, A. Hoefel, and myself [69]. This package, written for *Macaulay 2*, provides a suite of functions to experiment with edge and cover ideals. Many of the results discussed in the notes have been implemented into this package. Hopefully, the tools introduced in this tutorial will be the basis of your own research results!

As a final note, although I primarily discuss the `EdgeIdeal` package, I would recommend that your also become familiar with the packages `SimplicialComplexes`, written by S. Popescu, G.G. Smith, and M. Stillman (see [108]), and `SimplicialDecomposability` by D.W. Cook II (see [42]). The first package contains a number of useful functions related to simplicial complexes. In fact, the `EdgeIdeals` package requires a number of functions from this package. The `SimplicialDecomposability` package of D.W. Cook II is useful if you wish to study the properties of the simplicial complex associated to the edge or cover ideal of a graph.

A.V. Tuyl (✉)
Department of Mathematical Sciences, Lakehead University, Thunder Bay, Canada, ON P7B 5E1
e-mail: avantuyl@lakeheadu.ca

A.M. Bigatti et al. (eds.), *Monomial Ideals, Computations and Applications*,
Lecture Notes in Mathematics 2083, DOI 10.1007/978-3-642-38742-5_4,
© Springer-Verlag Berlin Heidelberg 2013

1.1   Getting Started

Obviously, the first thing you need to do is install the latest version[1] of *Macaulay 2* on your computer. The download page is here:

http://www.math.uiuc.edu/Macaulay2/Downloads/

Pick the appropriate operating system, and then follow the instructions. This may take some time and patience.

I am going to assume that you have installed *Macaulay 2* and now have it working. To familiarize yourself with the basic syntax and some simple examples, a good place to start is this web page:

http://www.math.uiuc.edu/Macaulay2/GettingStarted/

If you have never used *Macaulay 2*, take a couple of minutes to try a couple of the sample sessions.

1.2   The EdgeIdeals Package

Now that you have *Macaulay 2* installed, we want to load the EdgeIdeals package. If you are using a current version of *Macaulay 2* (i.e., a version $\geq$ 1.2), then this package should already be included with your installation of *Macaulay 2*, and it simply has to be installed.

Remark 1. If you have an older version, or if your version does not include this package, you should first download the source code from this link:

http://j-sag.org/Volume1/EdgeIdeals.m2

Save the code in a file named EdgeIdeals.m2, and save the file into your working directory. You can now return to the directions below. Note that when you run the command installPackage ``EdgeIdeals'', *Macaulay 2* will install the package where it can always find it in the future.

Open *Macaulay 2* and input the following command

```
Macaulay2, version 1.4
with packages: ConwayPolynomials, Elimination, IntegralClosure, LLLBases,
               PrimaryDecomposition, ReesAlgebra, TangentCone

i1 : installPackage "EdgeIdeals"
```

---

[1] At the time of writing this tutorial, the current version was 1.4.

You will only need to enter this command the first time you use the package. In the background, this command is making all the help pages. Once you have installed the package, you do not need to use the command again, but instead, use the instructions below. If you wish, you can start a new session by typing `restart`.

When we first start *Macaulay 2*, we start with following screen:

```
Macaulay2, version 1.4
with packages: ConwayPolynomials, Elimination, IntegralClosure, LLLBases,
               PrimaryDecomposition, ReesAlgebra, TangentCone

i1 :
```

At the prompt, type the following command to load the package `EdgeIdeals`:

```
i1 : loadPackage "EdgeIdeals"

o1 = EdgeIdeals

o1 : Package

i2 : loadedPackages

o2 = {EdgeIdeals, SimplicialComplexes, SimpleDoc, Elimination, LLLBases,
      --------------------------------------------------------------------
      IntegralClosure, PrimaryDecomposition, Classic, TangentCone,
      --------------------------------------------------------------------
      ReesAlgebra, ConwayPolynomials, Core}

o2 : List
```

The second command returns all the packages currently loaded in *Macaulay 2*. Note that not only is the `EdgeIdeals` package loaded, but so is the `SimplicialComplexes` package. Many of the functions in `EdgeIdeals` run "on top" of `SimplicialComplexes`.

We are now ready to try out `EdgeIdeals`. To get going, we spend a little time discussing how to input a finite simple graph. As a concrete example, suppose that we want to study the graph

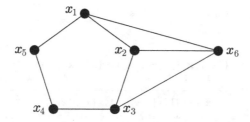

We enter this information in such a way that *Macaulay 2* recognizes it as a graph. There are a couple of ways to do this. The first way is to input a polynomial ring to denote the vertices, and then represent the edges as a list. For example

```
i3 : R = QQ[x_1..x_6]

o3 = R

o3 : PolynomialRing

i4 : E = {{x_1,x_2},{x_2,x_3},{x_3,x_4},{x_4,x_5},{x_5,x_1},{x_1,x_6},
          {x_2,x_6},{x_3,x_6}}

o4 = {{x , x }, {x , x }, {x , x }, {x , x }, {x , x }, {x , x }, {x , x },
         1   2     2   3     3   4     4   5     5   1     1   6     2   6
     ----------------------------------------------------------------------
     {x , x }}
       3   6

o4 : List

i5 : H = graph(R,E)

o5 = Graph{edges => {{x , x }, {x , x }, {x , x }, {x , x }, {x , x }, }
                       1   2     2   3     3   4     4   5     5   1
           -----------------------------------------------------------
           {x , x }, {x , x }, {x , x }}
             1   6     2   6     3   6
           ring => R
           vertices => {x , x , x , x , x , x }
                         1   2   3   4   5   6

o5 : Graph
```

Alternatively, the edges can be represented as the generators of a square-free quadratic monomial ideal. If no ring is passed to the command graph, it takes the variables of the current ring as the vertices of the graph. As an example, here is an alternative way to input the above graph into *Macaulay 2*:

```
i6 : e = monomialIdeal(x_1*x_2,x_2*x_3,x_3*x_4,x_4*x_5,x_5*x_1,x_1*x_6,
                       x_2*x_6,x_3*x_6)

o6 = monomialIdeal (x x , x x , x x , x x , x x , x x , x x , x x )
                     1 2   2 3   3 4   1 5   4 5   1 6   2 6   3 6

o6 : MonomialIdeal of R

i7 : G = graph e

o7 = Graph{edges => {{x , x }, {x , x }, {x , x }, {x , x }, {x , x }, }
                       1   2     2   3     3   4     1   5     4   5
           ------------------------------------------------------------
           {x , x }, {x , x }, {x , x }}
             1   6     2   6     3   6
           ring => R
           vertices => {x , x , x , x , x , x }
                         1   2   3   4   5   6
```

```
o7 : Graph

i8 : G==H

o8 = true
```

Now that we have an object called a `Graph`, we can ask about its edge and cover
ideals. Both of these ideals can be easily obtained using the following commands:

```
i9 : i = edgeIdeal G

o6 = monomialIdeal (x x , x x , x x , x x , x x , x x , x x , x x )
                     1 2    2 3    3 4    1 5    4 5    1 6    2 6    3 6

o9 : MonomialIdeal of R

i10 : j = coverIdeal G

o10 = monomialIdeal (x x x x , x x x x , x x x x , x x x x , x x x x ,
                      1 2 3 4   1 2 3 5   1 2 4 6   1 3 4 6   1 3 5 6
      -------------------------------------------------------------
      x x x x , x x x x )
       2 3 5 6   2 4 5 6

o10 : MonomialIdeal of R
```

The generators of $J(G)$ are the minimal vertex covers of G; convince yourself
that the generators given in the above example are indeed the minimal vertex covers
of the graph.

Recall that we showed that the Alexander dual of the edge ideal $I(G)$ equals the
cover ideal of $J(G)$. We can verify this for this ideal using a command from the
`SimplicialComplexes` package (which is also loaded):

```
i11 : dual i == j

o11 = true
```

Once you have inputted your graph, you can now compute some of its graph
theoretic invariants. For example, the chromatic number of the graph is computed as

```
i12 : chromaticNumber G

o12 = 3
```

To compute this number, we use the fact that

$$\chi(G) = \min\{d \mid (x_1 \cdots x_n)^{d-1} \in J(G)^d\}$$

as proved in Sect. 1 of chapter "A Beginner's Guide to Edge and Cover Ideals".
Similarly, Fröberg's Theorem gives us an algebraic characterization of chordal
graphs. We can therefore check if G is chordal:

```
i13 : isChordal G

o13 = false
```

To facilitate experimentation, we have built a number of functions to create commonly occurring graphs, like cycles and cliques. Here are some examples:

```
i14 : C6 = cycle R

o14 = Graph{edges => {{x , x }, {x , x }, {x , x }, {x , x }, {x , x },  }
                        1   2     2   3     3   4     4   5     5   6
            --------------------------------------------------
                      {x , x }}
                        1   6
            ring => R
            vertices => {x , x , x , x , x , x }
                          1   2   3   4   5   6

o14 : Graph

i15 : C5 = cycle(R,5)

o15 = Graph{edges => {{x , x }, {x , x }, {x , x }, {x , x }, {x , x }}}
                        1   2     2   3     3   4     4   5     1   5
            ring => R
            vertices => {x , x , x , x , x , x }
                          1   2   3   4   5   6

o15 : Graph
```

The command `cycle` will return a cycle of length equal to the number of variables in the ring R as a default. If a number n is given, it will make a cycle of that length using the first n variables. Cliques of size n are defined similarly:

```
i16 : K4 = completeGraph(R,4)

o16 = Graph{edges => {{x , x }, {x , x }, {x , x }, {x , x }, {x , x },  }
                        1   2     1   3     1   4     2   3     2   4
            --------------------------------------------------
                      {x , x }}
                        3   4
            ring => R
            vertices => {x , x , x , x , x , x }
                          1   2   3   4   5   6

o16 : Graph
```

The command `antiCycle` is similar in that it returns the graph of the complement of a cycle.

Also built into the `EdgeIdeals` package is a number of commands to construct subgraphs. For example, suppose that we wish to look at the induced subgraph of G on the vertices $P = \{x_1, x_2, x_6, x_5\}$. This can be done as follows:

```
i17 : P = {x_1,x_2,x_6,x_5}

o17 = {x , x , x , x }
        1   2   6   5

o17 : List

i18 : GP = inducedGraph(G,P)

o18 = Graph{edges => {{x , x }, {x , x }, {x , x }, {x , x }}}
                        1   2     1   5     1   6     2   6
            ring => QQ[x , x , x , x ]
                        1   2   6   5
            vertices => {x , x , x , x }
                          1   2   6   5

o18 : Graph
```

Another similar command that may prove helpful is deleteEdges which removes a collection of edges from a graph.

To facilitate research, the EdgeIdeals package includes a function called randomGraph. This function allows you to generate a random graph on defined number of vertices and edges, and is useful when creating conjectures. Here is an example of the this function in action:

```
i19 : randomGraph(R,8)

o19 = Graph{edges => {{x , x }, {x , x }, {x , x }, {x , x }, {x , x },  }
                        1   2     2   3     2   4     5   6     4   6
            --------------------------------------------------------------
            {x , x }, {x , x }, {x , x }}
              3   6     2   5     2   6
            ring => R
            vertices => {x , x , x , x , x , x }
                          1   2   3   4   5   6

o19 : Graph
```

In this case, we are asking for a random graph on six vertices (the number of variables in the polynomial ring R) with eight edges. This function can be used to test a large number of examples quickly.

As a final note, the documentation of the EdgeIdeals package can be found here:

http://www.math.uiuc.edu/Macaulay2/doc/Macaulay2-1.4/share/doc/Macaulay2/
EdgeIdeals/html/index.html

All the commands given in the package are listed on this page. Detailed documentation and examples can be found by clicking on the appropriate links.

2 Tutorials

I have included two tutorials to give you a chance to play around and experiment with edge and cover ideals using *Macaulay 2*. These tutorials were first given to the participants of MONICA. When required, the tutorials provide needed definitions, results, and references. Some of the initial problems ask you to prove some simple results in order to give you a feeling for the material, while other problems ask you to program some simple procedures using *Macaulay 2* in order to help you develop your *Macaulay 2* skills. The last batch of questions for each tutorial is a series of open questions. These questions are denoted by an asterisk. (If you come up with any ideas, I would love to hear them!)

2.1 Tutorial 1: Splitting Monomial Ideals

In this tutorial, we explore some of the properties of splitting monomial ideals as discussed in Sect. 2 of chapter "A Beginner's Guide to Edge and Cover Ideals".

Exercise 2. Suppose $I = J + K$ is a Betti splitting. Prove that

$$\text{reg}(I) = \max\{\text{reg}(J), \text{reg}(K), \text{reg}(J \cap K) - 1\}.$$

Here, $\text{reg}(-)$ denotes the regularity of the given ideal.

Remark. This result can be quite useful when doing induction. For example, this fact was used to give a new proof for the regularity of the edge ideal of a tree [83].

Exercise 3. Write a *Macaulay 2* program that takes as input two monomial ideals J and K, and will return `true` or `false` depending upon whether $J + K$ is a Betti splitting.

Hint 1. *The command* `betti res I` *will return the Betti diagram of the ideal I. Read through the* `betti` *documentation in order to extract out the information you are looking for. If you are interested in a particular graded Betti number, you may wish to first define the function:*

```
beta = (i,j,I) -> (betti res I)#(i,{j},j)
```

Exercise 4 (Importance of char(k)). Consider the following ideal in $R = k[x_1, \ldots, x_6]$:

$$I = (x_1x_2x_4, \ x_1x_2x_6, \ x_1x_3x_5, \ x_1x_3x_4, \ x_1x_5x_6,$$

$$x_2x_4x_5, \ x_2x_3x_6, \ x_2x_3x_5, \ x_3x_4x_6, \ x_4x_5x_6).$$

Fix a variable x_i, and form an x_i-partition of I, i.e., let J be the ideal generated by all the generators of I divisible by x_i, and let K be the ideal generated by the remaining generators. Use *Macaulay 2* to show $I = J + K$ is a Betti splitting in $\text{char}(k) = 2$, but not a Betti splitting if $\text{char}(k) \neq 2$.

Hint 2. *One way to input a ring of characteristic two is*

```
i1 : S = ZZ/(2)[a,b,c]
```

Definition 5. Let $I(G)$ be the edge ideal of a graph. For any edge $e = \{x_i, x_j\}$, we have the partition

$$I(G) = \langle x_i x_j \rangle + I(G \setminus e)$$

where $G \setminus e$ is the graph G with the edge e removed. We call e a *splitting edge* if this partition is a Betti splitting.

Exercise 6. Consider the graph

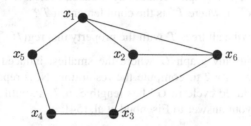

Determine which edges of this graph are splitting edges.

Exercise 7. Write a program in *Macaulay 2* that inputs a graph, and returns all the edges in the graph that are splitting edges.

Exercise 8. Let $G = C_n$ be a cycle of length $n \geq 4$. Prove that G has no splitting edge.

Exercise 9. Let $G = K_n$ be the clique of size $n \geq 3$. Prove that every edge of G is a splitting edge.

Exercise 10. Find a graph G that is not a cycle, but no edge is a splitting edge, or prove that this is not possible. Then, find a graph G that is not a clique, but every edge is a splitting edge, or prove that this is not possible.

Exercise 11. A vertex v is called a *leaf* if $\deg v = 1$. Suppose that v is a leaf, and $e = \{v, u\}$ is the only edge that contains v. Prove that e is a splitting edge.

Exercise 12. Let $N(x) = \{y \mid \{x, y\} \in E(G)\}$ be the neighbours of x. Make a conjecture about $\{x, y\}$ being a splitting edge in terms of $N(x) \cup N(y)$. Compare your answer to [83].

$\star$**Exercise 13.** *Is the number of splitting edges related to any invariants of G or $I(G)$?*

⋆**Exercise 14.** *Find other ways to split $I(G)$.*

⋆**Exercise 15.** *Are there any nice ways to construct Betti splittings of the cover ideal $J(G)$? (I am only aware of how to split $J(G)$ in the case that $R/J(G)$ is Cohen–Macaulay and G is bipartite [65].)*

⋆**Exercise 16.** *Are there Betti splittings of the ideals $I(G)^s$ and $J(G)^s$, for some integer s?*

2.2 Tutorial 2: Regularity

In this tutorial, we look at the regularity of edge and cover ideals.

Exercise 17. A *tree* is a graph without any induced cycles. If T is a tree, what is the regularity of $I(T^c)$, where T^c is the complement of T?

Exercise 18. Describe all trees T with the property that $\text{reg}(I(T)) = 2$.

Exercise 19. Create any graph G where the smallest induced cycle of G^c has length 4. Use *Macaulay 2* to compute the resolution. Now repeat for a graph G whose smallest induced cycle in G^c has length $5, 6, 7, \dots$ until you observe your pattern. Compare your answer to Eisenbud et al. [54].

Exercise 20. If you would like to see the code of a *Macaulay 2* function, you can use

```
code methods use ⟨ function name ⟩
```

Look at the code for `smallestCycleSize`. Try to figure out how *Macaulay 2* finds the smallest induced cycle in a graph.

Exercise 21. Write a *Macaulay 2* function that checks if a graph has an induced four cycle.

Hint 3. *Use the fact that*

$$\beta_{1,4}(I(G)) = c_4(G^c)$$

where $c_4(H)$ denotes the number of induced four cycles in the graph H (see [193]).

Exercise 22. Write a *Macaulay 2* function that tests whether an ideal has a linear resolution.

Exercise 23. Nevo and Peeva [151] have made the following conjecture:

Conjecture 24. For all graphs G, if G^c has no induced four cycles, then there exists an integer s such that $I(G)^s$ has a linear resolution.

Using the command `randomGraph`, find ten graphs where the conjecture is true, and for each graph, find the smallest integer s where $I(G)^s$ has a linear resolution.

Exercise 25. The *path* of length n, denoted P_n is the graph with vertex set $\{x_1, x_2, \ldots, x_n\}$ and edge set

$$\{\{x_1, x_2\}, \{x_2, x_3\}, \ldots, \{x_{n-1}, x_n\}\}.$$

Compute the regularity of $I(P_n)$ for some n until you find a pattern. Compare your result to Jacques [114].

⋆Exercise 26. *Let T be a tree. Find a formula for* $\mathrm{reg}(I(T)^s)$ *as s varies.*

Hint 4. *You may wish to start with the case that $T = P_n$ first.*

⋆Exercise 27. *Find a formula for* $\mathrm{reg}(J(G))$ *and* $\mathrm{reg}(I(G))$ *for any graph.*

Hint 5. *This problem is probably too open ended. I am not aware of many results on the regularity of $J(G)$. For edge ideals, more is known (do a Google search on "regularity edge ideals"). For bipartite graphs, we almost have a complete story. See [190] for more.*

Part III
Local Cohomology

Local Cohomology Modules Supported on Monomial Ideals

Josep Àlvarez Montaner

Introduction

Local cohomology was introduced by A. Grothendieck in the early 1960s and quickly became an indispensable tool in Commutative Algebra. Despite the effort of many authors in the study of these modules, their structure is still quite unknown. C. Huneke [109] raised some basic questions concerning local cohomology:

- Annihilation of local cohomology modules.
- Finitely generation of local cohomology modules.
- Artinianity of local cohomology modules.
- Finiteness of the associated primes set of local cohomology modules.

In general, one can not even say when they vanish. Moreover, when they do not vanish they are rarely finitely generated. However, in some situations these modules verify some finiteness properties that provide a better understanding of their structure.

Josep Àlvarez Montaner was partially supported by SGR2009-1284 and MTM2010-20279-C02-01.

J. Àlvarez Montaner (✉)
Departamento de Matemàtica Aplicada I
Universitat Politècnica de Catalunya
Av. Diagonal 647, Barcelona 08028, Spain
e-mail: Josep.Alvarez@upc.edu

A.M. Bigatti et al. (eds.), *Monomial Ideals, Computations and Applications*, 109
Lecture Notes in Mathematics 2083, DOI 10.1007/978-3-642-38742-5_5,
© Springer-Verlag Berlin Heidelberg 2013

For example, when we have a regular local ring containing a field, the set of associated primes of local cohomology modules is finite. Moreover, the Bass numbers of these modules with respect to any prime ideal are also finite. This result has been proved by C. Huneke and R.Y. Sharp [110] (see also [133]) in positive characteristic and by G. Lyubeznik in the zero characteristic case [132].

The key idea behind these results is that we can enrich our local cohomology modules with an extra structure that makes easier to describe them (see [134] for a nice survey). In characteristic zero we have that local cohomology modules are finitely generated modules over the ring of differential operators, in fact they are holonomic. In positive characteristic we have that local cohomology modules are F-finite F-modules. This is a notion developed by G. Lyubeznik in [133] with the help of the Frobenius morphism. It is proved in [133] that this class of modules satisfy analogous properties to those of holonomic D-modules.

In this survey we turn our attention to the case of local cohomology modules supported on monomial ideals. The interest for this kind of modules is not new since it provides a nice family of examples that are easier to deal with when one considers some of the basic problems concerning local cohomology but also because their graded pieces can be used to describe cohomology groups of sheaves on a toric variety (see [55, 147]).

A lot of progress in the study of these modules has been made in recent years based on the fact that they have a structure as $\mathbb{Z}^n$-graded module. We will give an overview of this approach but we will put more emphasis on the D-module approach because of the habit of the author but also because it gives a new perspective for the reader that has a stronger background in Combinatorial Commutative Algebra.

In the first part of Sect. 1 we will review the basics on the theory of local cohomology modules we will use throughout this work. The second part of Sect. 1 is devoted to the theory of D-modules. We start from scratch with the basic definitions and we illustrate them with the main examples we will consider in this work: the polynomial ring, localizations of this polynomial ring and local cohomology modules. Then we quickly direct ourselves to the construction of the *characteristic cycle*. This is an invariant that gives us a lot of information about our module. Again, we will illustrate this fact with some basic examples.

In Sect. 2 we will study the structure of local cohomology modules with support a monomial ideal from two different point of views. First we will highlight the main results obtained in [146, 187, 200] where the $\mathbb{Z}^n$-graded structure of these modules is described. Then we will consider the D-module approach given in [3, 7, 8]. We tried to illustrate this approach with a lot of examples since we assume that the reader is not familiar with this point of view. In the end, it turns out that both approaches are equivalent and one can build a dictionary between the $\mathbb{Z}^n$-graded and the D-module structure of these local cohomology modules. More striking is the dictionary we will build between local cohomology modules and free resolutions and we will extend in Sect. 3.

In Sect. 3 we turn our attention to Bass numbers of local cohomology modules with support a monomial ideal. A formula to describe these invariants using the $\mathbb{Z}^n$-graded structure has been given by K. Yanagawa in [200]. From the D-modules point of view we have to refer to the algorithms given by the author in [3, 5]. However, in this survey we will mainly report some recent work by A. Vahidi and the author [10]. The approach given in [10] is to consider the pieces of the composition of local cohomology modules. It turns out that the structure of these local cohomology modules (as $\mathbb{Z}^n$-graded module or module with variation zero) is required to compute their Bass numbers. Using the dictionary between local cohomology modules and free resolutions, one may understand the Lyubeznik numbers as a measure for the acyclicity of the linear strands of the free resolution of the Alexander dual ideal.

Bass numbers of local cohomology modules do not behave as nicely as in the case of finitely generated modules but we can control them using their structure of the local cohomology modules. This control leads to some partial description of its injective resolution. In particular we give a bound for the injective dimension in terms of the small support of these modules.

We also include an appendix where we fix the notations and some basic facts on $\mathbb{Z}^n$-graded free and injective resolutions. We recommend the reader to take a look at this appendix whenever one finds some unexplained reference to these concepts.

In chapter "Local Cohomology Using Macaulay2" we provide some functions developed with O. Fernández-Ramos using `Macaulay 2` that allows us to compute the characteristic cycle of local cohomology modules supported on monomial ideals and also Lyubeznik numbers. We also included a Tutorial aimed to start experimenting with an small sample of exercises. Once this set of exercises is completed we encourage the reader to develop with their own examples and come out with their own results.

1 Basics on Local Cohomology Modules

1.1 Basic Definitions

We will start with a quick introduction to the theory of local cohomology modules as introduced in [81]. Our main reference will be the book of M.P. Brodmann and R.Y. Sharp [29] but we also recommend to take a look at [52, 113]. Then we will turn our attention to some finiteness results on these modules that prompted G. Lyubeznik [132] to define a new set of invariants that we will study in Sect. 3.

Let $I \subseteq R$ be an ideal of a commutative Noetherian ring R. The usual way to introduce local cohomology modules is through *the I-torsion functor* over the category of R-modules. This is a functor $\Gamma_I : \mathrm{Mod}(R) \longrightarrow \mathrm{Mod}(R)$ defined for any R-module M as:

$$\Gamma_I(M) := \{x \in M \mid I^n x = 0 \text{ for some } n \geq 1\}.$$

The category $\mathrm{Mod}(R)$ of R-modules is abelian and has enough injectives. On the other hand, the I-torsion functor Γ_I is additive, covariant and left-exact so it makes sense to consider the right derived functors of Γ_I. These are called *the local cohomology modules of M with respect to I* and are denoted by

$$H_I^r(M) := \mathbb{R}^r \Gamma_I(M).$$

Remark 1. The functor of I-torsion can be expressed as

$$\Gamma_I(M) = \bigcup_{n \geq 0} (0 :_M I^n) = \varinjlim_{n \in \mathbb{N}} \mathrm{Hom}_R(R/I^n, M),$$

so we have $H_I^r(M) = \varinjlim_{n \in \mathbb{N}} \mathrm{Ext}_R^r(R/I^n, M)$.

Throughout this work, we will prefer to describe these local cohomology modules by means of the *Čech complex*. It will help us to put our hands on these objects and make some explicit computations but it will be also helpful when we want to enrich the local cohomology modules with an extra structure as a D-module structure, a Frobenius action structure or, in the case of monomial ideals, a $\mathbb{Z}^n$-graded structure.

Assume that our ideal $I \subseteq R$ is generated by elements $f_1, \ldots, f_s \in R$. For any R-module M, the Čech complex of M with respect to I, that we will denote $\check{C}_I^\bullet(M)$ is the following complex:

$$0 \longrightarrow M \xrightarrow{d_0} \bigoplus_{1 \leq i_1 \leq s} M_{f_{i_1}} \xrightarrow{d_1} \cdots \xrightarrow{d_{p-1}} \bigoplus_{1 \leq i_1 < \cdots < i_p \leq s} M_{f_{i_1} \cdots f_{i_p}} \xrightarrow{d_p} \cdots \xrightarrow{d_{s-1}} M_{f_1 \cdots f_s} \longrightarrow 0,$$

where the differentials d_p are defined by using the canonical localization morphism on every component $M_{f_{i_1} \cdots f_{i_p}} \longrightarrow M_{f_{j_1} \cdots f_{j_{p+1}}}$ as follows:

$$d_p(m) = \begin{cases} (-1)^k \dfrac{m}{1} & \text{if } \{i_1, \ldots, i_p\} = \{j_1, \ldots, \widehat{j_k}, \ldots, j_{p+1}\}, \\ 0 & \text{otherwise.} \end{cases}$$

Then, the local cohomology modules are nothing but the cohomology modules of this complex, i.e.

$$H_I^r(M) = H^r(\check{C}_I^\bullet(M)) = \ker d_r / \mathrm{Im}\, d_{r-1}.$$

The following basic properties will be often used without further mention. The main one states that local cohomology only depends on the radical of the ideal.

- $H_I^r(M) = H_{\mathrm{rad}(I)}^r(M)$, for all $r \geq 0$.
- Let $\{M_j\}_{j \in J}$ be an inductive system of R-modules. Then:

$$H_I^r(\varinjlim M_j) = \varinjlim H_I^r(M_j).$$

- *Invariance with respect to base ring:* Let $R \longrightarrow S$ be a homomorphism of rings. Let $I \subseteq R$ be an ideal and M an S-module. Then:

$$H_{IS}^r(M) \cong H_I^r(M).$$

- *Flat base change:* Let $R \longrightarrow S$ be a flat homomorphism of rings. Let $I \subseteq R$ be an ideal and M an R-module. Then:

$$H_{IS}^r(M \otimes_R S) \cong H_I^r(M) \otimes_R S.$$

Local cohomology modules are in general not finitely generated as R-modules, so they are difficult to deal with. In order to extract some properties of these modules the usual method is to use several exact sequences or spectral sequences involving these modules. Enumerated are some examples we will use in this work, for details we refer to [29].

- *Long exact sequence of local cohomology:* Let $I \subseteq R$ be an ideal and $0 \longrightarrow M_1 \longrightarrow M_2 \longrightarrow M_3 \longrightarrow 0$ an exact sequence of R-modules. Then we have the exact sequence:

$$\cdots \longrightarrow H_I^r(M_1) \longrightarrow H_I^r(M_2) \longrightarrow H_I^r(M_3) \longrightarrow H_I^{r+1}(M_1) \longrightarrow \cdots$$

- *Mayer–Vietoris long exact sequence:* Let $I, J \subseteq R$ be ideals and M an R-module. Then we have the exact sequence:

$$\cdots \longrightarrow H_{I+J}^r(M) \longrightarrow H_I^r(M) \oplus H_J^r(M) \longrightarrow H_{I \cap J}^r(M) \longrightarrow H_{I+J}^{r+1}(M) \longrightarrow \cdots$$

- *Brodmann's long exact sequence:* Let $I \subseteq R$ be an ideal and M an R-module. For any element $f \in R$ we have the exact sequence:

$$\cdots \longrightarrow H_{I+(f)}^r(M) \longrightarrow H_I^r(M) \longrightarrow H_I^r(M)_f \longrightarrow H_{I+(f)}^{r+1}(M) \longrightarrow \cdots$$

- *Grothendieck's spectral sequence:* Let $I, J \subseteq R$ be ideals and M an R-module. Then we have the spectral sequence:

$$E_2^{p,q} = H_J^p(H_I^q(M)) \Longrightarrow H_{I+J}^{p+q}(M).$$

Quite recently, a generalization of the Mayer–Vietoris sequence has been given in [7]. We give the presentation[1] introduced in [135] since it requires less notation.

---

[1]The presentation given in [135] slightly differs from the one given in [7] at the E_1-page but they coincide at the E_2-page.

- *Mayer–Vietoris spectral sequence:* Let $I = I_1 \cap \cdots \cap I_m$ be a decomposition of an ideal $I \subseteq R$ and M an R-module. Then we have the spectral sequence:

$$E_1^{-p,q} = \bigoplus_{1 \leq i_1 < \cdots < i_p \leq m} H_{I_{i_1} + \cdots + I_{i_p}}^q (M) \Longrightarrow H_I^{q-p}(M).$$

1.2 Finiteness Results

Although local cohomology modules are in general not finitely generated, under certain conditions they satisfy some finiteness properties that provide a better understanding of their structure. A turning point in the theory came at the beginning of the 1990s with the following remarkable result.

Let R be any regular ring containing a field of characteristic zero and let $I \subseteq R$ be an ideal. Then, local cohomology modules $H_I^r(R)$ satisfy the following properties:

1. $H_m^p(H_I^r(R))$ is injective, where m is any maximal ideal of R.
2. $\mathrm{id}_R(H_I^r(R)) \leq \dim_R \mathrm{Supp}_R H_I^r(R)$.
3. The set of the associated primes of $H_I^r(R)$ is finite.
4. All the Bass numbers of $H_I^r(R)$ are finite.

This result was proved by C. Huneke and R.Y. Sharp [110] for regular rings containing a field of positive characteristic using the Frobenius map. In his paper [132], G. Lyubeznik proved the case of regular rings containing a field of characteristic zero using the theory of algebraic D-modules. Even though local cohomology modules had been already used in the theory of D-modules, the work of G. Lyubeznik became the first application of this theory to an explicit problem in Commutative Algebra. The main point is that local cohomology modules $H_I^i(R)$ are finitely generated as D-modules. In fact they are *holonomic* D-modules. This class of modules form an abelian subcategory of the category of D-modules satisfying some good properties, in particular they have finite length.

The results of G. Lyubeznik are slightly more general than those of C. Huneke and R.Y. Sharp. To prove the validity of this generalization in positive characteristic he introduced the theory of F-modules in [133]. In particular he introduced the class of F-finite F-modules that satisfy analogous properties to those of holonomic D-modules. Local cohomology modules belong to this class so one can follow a similar program to prove the same results in positive characteristic.

1.3 Lyubeznik Numbers

Using the finiteness of Bass numbers G. Lyubeznik [132] defined a new set of numerical invariants. More precisely, let (R, m, k) be a regular local ring of dimension n containing a field k and A a local ring which admits a surjective ring homomorphism $\pi : R \longrightarrow A$. Set $I = \operatorname{Ker} \pi$, then we consider the Bass numbers

$$\lambda_{p,i}(A) := \mu_p(\mathfrak{m}, H_I^{n-i}(R)).$$

This invariants depend only on A, i and p, but neither on R nor on π. Completion does not change $\lambda_{p,i}(A)$ so one can assume $R = k[[x_1, \ldots, x_n]]$. It is worthwhile to point out that, by Lyubeznik [132, Lemma 1.4], we have

$$\lambda_{p,i}(R/I) = \mu_p(\mathfrak{m}, H_I^{n-i}(R)) = \mu_0(\mathfrak{m}, H_{\mathfrak{m}}^p(H_I^{n-i}(R))),$$

i.e. $H_{\mathfrak{m}}^p(H_I^{n-i}(R)) \cong E(R/\mathfrak{m})^{\lambda_{p,i}}$. This is the approach that we will use in Sect. 3.

Lyubeznik numbers satisfy $\lambda_{d,d}(A) \neq 0$ and $\lambda_{p,i}(A) = 0$ for $i > d$, $p > i$, where $d = \dim A$. Therefore we can collect them in what we refer as *Lyubeznik table*:

$$\Lambda(A) = \begin{pmatrix} \lambda_{0,0} & \cdots & \lambda_{0,d} \\ & \ddots & \vdots \\ & & \lambda_{d,d} \end{pmatrix}.$$

They have some interesting topological interpretation, as it was already pointed out in [132] but not so many examples can be found in the literature. The basic ones being the following:

Example 2. Assume that we only have one non-vanishing local cohomology module, i.e. $H_I^r(R) = 0$ for all $r \neq \operatorname{ht} I$. Then, using Grothendieck's spectral sequence to compute the composition $H_{\mathfrak{m}}^p(H_I^{n-i}(R))$, we obtain a trivial Lyubeznik table.

$$\Lambda(R/I) = \begin{pmatrix} 0 & \cdots & 0 \\ & \ddots & \vdots \\ & & 1 \end{pmatrix}.$$

This situation is achieved in the following cases:

- R/I is a complete intersection.
- R/I is Cohen–Macaulay and contains a field of positive characteristic.
- R/I is Cohen–Macaulay and I is a squarefree monomial ideal.

Example 3. Assume that $\operatorname{Supp}_R H_I^r(R) \subseteq V(\mathfrak{m})$ for all $r \neq \operatorname{ht} I$. Rings satisfying this property can be viewed as rings which behave cohomologically like an isolated singularity. Then, the corresponding Lyubeznik table has the following shape:

$$
\Lambda(R/I) = \begin{pmatrix} 0 & \lambda_{0,1} & \lambda_{0,2} & \cdots & \lambda_{0,d-1} & 0 \\ & 0 & 0 & \cdots & 0 & 0 \\ & & 0 & \cdots & 0 & \lambda_{0,d-1} \\ & & & \ddots & \vdots & \vdots \\ & & & & 0 & \lambda_{0,2} \\ & & & & & \lambda_{0,1}+1 \end{pmatrix}.
$$

That is $\lambda_{a,d} - \delta_{a,d} = \lambda_{0,d-a+1}$ for $2 \leq a \leq d$, where $\delta_{a,d}$ is the Kronecker delta function. This result was proved in [74] for isolated singularities over $\mathbb{C}$. It was then generalized in any characteristic in [23,24] (see also [170]).

1.4   The Theory of D-Modules

In this section we will provide some basic foundations in the theory of modules over the ring of differential operators that we will use throughout this work. We will only scratch the surface of this theory so we encourage the interested reader to take a deeper look at the available literature. The main references that we will use are [21,26,45].

Let k be a subring of a commutative Noetherian ring R. The ring of differential operators D_R, introduced by A. Grothendieck in [82, §16.8], is the subring of $\mathrm{End}_k(R)$ generated by the k-linear derivations and the multiplications by elements of R. In these lectures we will mainly consider the case where R is a polynomial ring over a field k of characteristic zero.

1.5   The Weyl Algebra

Let $R = k[x_1, \ldots, x_n]$ be a polynomial ring over a field k of characteristic zero. The ring of differential operators D_R coincides with the Weyl algebra $A_n(k) := k[x_1, \ldots, x_n]\langle \partial_1, \ldots, \partial_n \rangle$, i.e. the non commutative R-algebra generated by the partial derivatives $\partial_i = \frac{d}{dx_i}$, with the relations given by:

$$x_i x_j - x_j x_i = 0,$$
$$\partial_i \partial_j - \partial_j \partial_i = 0,$$
$$\partial_i r - r \partial_i = \frac{dr}{dx_i} \text{ where } r \in R.$$

Any element $P \in D_R$ can be uniquely written as a finite sum

$$
P = \sum a_{\alpha\beta} \mathbf{x}^\alpha \partial^\beta,
$$

where $a_{\alpha\beta} \in k$ and we use the notation $\mathbf{x}^\alpha := x_1^{\alpha_1} \cdots x_n^{\alpha_n}$, $\partial^\beta := \partial_1^{\beta_1} \cdots \partial_n^{\beta_n}$ for any $\alpha, \beta \in \mathbb{N}^n$. As usual, we will denote $|\alpha| = \alpha_1 + \cdots + \alpha_n$.

The ring of differential operators D_R is a left and right Noetherian ring and has an increasing filtration $\{\Sigma_v\}_{v \geq 0}$ of finitely generated R-submodules satisfying $\forall v, w \geq 0$:

- $\bigcup_{v \in \mathbb{N}} \Sigma_v = D_R$

- $\Sigma_v \Sigma_w = \Sigma_{v+w}.$

such that the corresponding associated graded ring

$$gr_\Sigma D_R = \Sigma_0 \oplus \frac{\Sigma_1}{\Sigma_0} \oplus \cdots$$

is isomorphic to a polynomial ring in $2n$ variables $k[x_1, \ldots, x_n, \xi_1, \ldots, \xi_n]$.

There are several ways to construct such filtration but we will only consider the natural increasing filtration given by the order. Recall that the *order* of a differential operator $P = \sum a_{\alpha\beta} \mathbf{x}^\alpha \partial^\beta$ is the integer

$$o(P) = \sup\{|\beta| \mid a_{\alpha\beta} \neq 0\}.$$

Then, the *order filtration* is given by the sets of differential operators of order less than v,

$$\Sigma_v^{(0,1)} = \{P \in D_R \mid o(P) \leq v\}.$$

The superscript we use is just because this filtration is associated to the weight vector[2] $(0, 1) \in \mathbb{Z}^{2n}$, i.e. we set $\deg x_i = 0$ and $\deg \partial_i = 1$. For simplicity we will also denote the associated graded ring by $gr_{(0,1)} D_R$ and an explicit isomorphism with a polynomial ring in $2n$ variables is given as follows:

The *principal symbol* of a differential operator $P = \sum a_{\alpha\beta} \mathbf{x}^\alpha \partial^\beta \in D_R$ is the element of the polynomial ring $k[x_1, \ldots, x_n, \xi_1, \ldots, \xi_n]$ in the independent variables $\xi_1, \ldots, \xi_n$ defined by:

$$\sigma_{(0,1)}(P) = \sum_{|\alpha|=o(P)} a_{\alpha\beta} \mathbf{x}^\alpha \xi^\beta,$$

where we use again the multidegree notation $\xi^\beta := \xi_1^{\beta_1} \cdots \xi_n^{\beta_n}$. Then, the map:

$$\begin{array}{ccc} gr_{(0,1)} D_R & \longrightarrow & k[x_1, \ldots, x_n, \xi_1, \ldots, \xi_n] \\ P & \cdots\cdots\cdots\rightarrow & \sigma_{(0,1)}(P) \end{array}$$

is an isomorphism of commutative rings.

---

[2]One may also consider filtrations associated to other weight vectors $(u, v) \in \mathbb{Z}^{2n}$ with $u + v \geq 0$, but then the corresponding associated graded ring $gr_{(u,v)} D_R$ is not necessarily a polynomial ring.

Warning: At some point in these lectures we will also have to consider the ring of differential operators over the formal power series ring $S = k[[x_1, \ldots, x_n]]$. In this case we have $D_S = D_R \otimes_R S = k[[x_1, \ldots, x_n]]\langle \partial_1, \ldots, \partial_n \rangle$, i.e. the non commutative S-algebra generated by the partial derivatives $\partial_i = \frac{d}{dx_i}$, with the same relations as given before. We can mimic what we did before to prove that the graded ring $gr_{(0,1)} D_S$ associated to the order filtration on D_S is isomorphic to $k[[x_1, \ldots, x_n]][\xi_1, \ldots, \xi_n]$.

1.6 Modules over the Ring of Differential Operators

D_R is a non-commutative ring so by a D_R-module we will always mean a left D_R-module. Now we will present the main examples we will consider in these lectures: the polynomial ring R, the localizations R_f at any element $f \in R$ and the local cohomology modules $H_I^r(R)$ where $I \subseteq R$ is any ideal.

• *Polynomial ring R:* The action of x_i on a polynomial $f \in R$ is just the multiplication by x_i. The action of ∂_i is the usual derivation with respect to the corresponding variable, i.e. $\partial_i \cdot f = \frac{df}{dx_i}$.

Notice that we have the presentation

$$R = \frac{D_R}{D_R(\partial_1, \ldots, \partial_n)} = \frac{k[x_1, \ldots, x_n]\langle \partial_1, \ldots, \partial_n \rangle}{(\partial_1, \ldots, \partial_n)}.$$

In particular, R is a finitely generated D_R-module.

• *Localizations R_f:* Consider the localization of R at any polynomial $f \in R$

$$R_f := \{ \frac{g}{f^n} \mid g \in R, \ n \geq 0 \}.$$

Again, the action of x_i on $\frac{g}{f^n} \in R_f$ is the multiplication and the action of ∂_i is given by Leibniz rule.

A deep result states that R_f is the D_R-module generated by $\frac{1}{f^\ell}$, where ℓ is the smallest integer root of the so-called *Bernstein–Sato* polynomial of f. We will skip the details on this theory since we will not use it in this work. We highlight from this result that we have a presentation

$$R_f = D_R \cdot \frac{1}{f^\ell} = \frac{D_R}{\text{Ann}_{D_R}(\frac{1}{f^\ell})}$$

thus R_f is a finitely generated D_R-module. More generally, given any D_R-module M, the localization $M_f = M \otimes_R R_f$ is also a D_R-module.

- *Localizations at monomials:* Let $f = x_1 \cdots x_p$, $p \leq n$ be a monomial. Then we can give a more precise description[3]:

$$R_{x_1 \cdots x_p} = \frac{D_R}{D_R(x_1 \partial_1 + 1, \ldots, x_p \partial_p + 1, \partial_{p+1}, \ldots, \partial_n)}.$$

- *Local cohomology modules:* Let M be a D_R-module and $I \subseteq R$ any ideal. Then, using the Čech complex $\check{C}_I^{\bullet}(M)$ we can give a D_R-module structure on the local cohomology modules $H_I^r(M)$.

The local cohomology of the polynomial ring R with respect to the homogeneous maximal ideal $\mathfrak{m}$ has the following presentation (see [132])

$$H_{\mathfrak{m}}^n(R) = \frac{D_R}{D_R(x_1, \ldots, x_n)}$$

so it is finitely generated. It is also known that for any homogeneous prime ideal $(x_1, \ldots, x_p)$, $p \leq n$ there is only a non-vanishing local cohomology module that has the presentation

$$H_{(x_1, \ldots, x_p)}^p(R) = \frac{D_R}{D_R(x_1, \ldots, x_p, \partial_{p+1}, \ldots, \partial_n)}.$$

In the following subsection we will see that in general, the local cohomology modules $H_I^r(R)$ and $H_{\mathfrak{m}}^p(H_I^r(R))$ are finitely generated as D_R-modules.

1.6.1 Good Filtrations

A finitely generated D_R-module M has a *good* filtration $\{\Gamma_k\}_{k \geq 0}$ compatible with the filtration $\{\Sigma_v\}_{v \geq 0}$ on D_R, i.e. M has an increasing sequence of finitely generated R-submodules $\Gamma_0 \subseteq \Gamma_1 \subseteq \cdots \subseteq M$ satisfying:

- $\bigcup \Gamma_k = M$,
- $\Sigma_v \Gamma_k \subseteq \Gamma_{v+k}$.

such that the associated graded module $gr_\Gamma M = \Gamma_0 \oplus \frac{\Gamma_1}{\Gamma_0} \oplus \cdots$ is a finitely generated $gr_\Sigma D_R$-module.

Again, there are several ways to find good filtrations on a finitely generated D_R-module. When we have a presentation $M = \frac{D_R}{L}$ the order filtration on D_R induces a good filtration on M such that the corresponding associated graded module is

---

[3]Given the relation $x_i \partial_i + 1 = 0$ one may interpret ∂_i as the fraction $\frac{1}{x_i}$ in the localization.

$$gr_{(0,1)}M = \frac{gr_{(0,1)}D_R}{gr_{(0,1)}L} = \frac{k[x_1,\ldots,x_n,\xi_1,\ldots,\xi_n]}{gr_{(0,1)}L}$$

where $gr_{(0,1)}L = (\sigma_{(0,1)}(P) \mid P \in L)$.

1.6.2 Holonomic D_R-Modules

Let M be a finitely generated D_R-module, then we can define its dimension and multiplicity by means of its graded module $gr_\Gamma M$ associated to a good filtration $\{\Gamma_k\}_{k\geq 0}$. Recall that $gr_\Gamma M$ is finitely generated as a module over the polynomial ring $gr_\Sigma D_R$ so we can use the theory of Hilbert functions to compute its dimension and multiplicity. Namely, let $(x,\xi) = (x_1,\ldots,x_n,\xi_1,\ldots,\xi_n) \in k[x_1,\ldots,x_n,\xi_1,\ldots,\xi_n]$ be the homogeneous maximal ideal. The Hilbert series of the graded module $gr_\Gamma M$:

$$H(gr_\Gamma M;t) = \sum_{j\geq 0} \dim_k [(x,\xi)^j gr_\Gamma M/(x,\xi)^{j+1} gr_\Gamma M] \ t^j,$$

is of the form $H(gr_\Gamma M;t) = q(t)/(1-t)^d$, where $q(t) \in \mathbb{Z}[t,t^{-1}]$ is such that $q(1) \neq 0$. The Krull dimension of $gr_\Gamma M$ is d and the multiplicity of $gr_\Gamma M$ is $q(1)$. These integers are independent of the good filtration on M and are called the dimension and the multiplicity of M. We will denote them $d(M)$ and $e(M)$ respectively. In the next section, we will use a geometric description of the dimension given by the so-called *characteristic variety*.

The following result is a deep theorem, proved by M. Sato, T. Kawai and M. Kashiwara in [168] (see also [137]), by using microlocal techniques. Later, O. Gabber [71] gave a purely algebraic proof:

Theorem (Bernstein's inequality). *Let M be a non-zero finitely generated D_R-module. Then $d(M) \geq n$.*

Now we single out the important class of D_R-modules having the minimal possible dimension.

Definition (Holonomicity). One says that a finitely generated D_R-module M is *holonomic* if $M = 0$ or $d(M) = n$.

The class of holonomic modules has many good properties. Among them we find:

- Holonomic modules form a full abelian subcategory of the category of D_R-modules. In particular if $0 \longrightarrow M_1 \longrightarrow M_2 \longrightarrow M_3 \longrightarrow 0$ is an exact sequence of D_R-modules, then M_2 is holonomic if and only if M_1 and M_3 are both holonomic.
- M is holonomic if and only if M has finite length as D_R-module.

- M is holonomic if and only if $\mathrm{Ext}^i_{D_R}(M, D_R) = 0$ for all $i \neq n$.

The polynomial ring R, the localizations R_f at any element $f \in R$ and the local cohomology modules $H^r_I(R)$ are holonomic D_R-modules. We will check out this fact in the next subsection using the characteristic variety. We also point out that these modules are in fact regular holonomic modules in the sense of Mebkhout [139].

1.7 The Characteristic Variety

Our aim is to associate to a finitely generated D_R-module M equipped with a good filtration $\{\Gamma_k\}_{k\geq 0}$ an invariant that provides a lot of information on this module. Since $gr_\Gamma M$ is a finitely generated $gr_\Sigma D_R$-module, where $gr_\Sigma D_R = k[x_1, \ldots, x_n, \xi_1, \ldots, \xi_n]$, we may construct the following:

- *Characteristic ideal:* Is the ideal in $k[x_1, \ldots, x_n, \xi_1, \ldots, \xi_n]$ given by:

$$J_\Sigma(M) := \mathrm{rad}\,(\mathrm{Ann}_{gr_\Sigma D_R}(gr_\Gamma M)).$$

The characteristic ideal depends on the filtration $\{\Sigma_v\}_{v\geq 0}$ but, once the filtration is fixed, $J_\Sigma(M)$ is independent of the good filtration on M.
- *Characteristic variety:* Is the closed algebraic set given by:

$$C_\Sigma(M) := V(J_\Sigma(M)) \subseteq \mathrm{Spec}\,(k[x_1, \cdots, x_n, \xi_1, \ldots, \xi_n]).$$

From now on we are only going to consider the ring of differential operators with the order filtration $\{\Sigma^{(0,1)}_v\}_{v\geq 0}$ so the characteristic ideal and the characteristic variety that we will use in this work will be denoted simply as

$$J(M) := \mathrm{rad}\,(\mathrm{Ann}_{gr_{(0,1)} D_R}(gr_\Gamma M)),$$
$$C(M) := V(J(M)).$$

If the reader is interested on the behavior of the characteristic variety $C_{(u,v)}(M)$ associated to the filtration given by a weight vector (u, v) we recommend to take a look at [25, 177]. An interesting feature is that the Krull dimension of the characteristic variety does not depend on the filtration. This provides a geometric description of the dimension of a finitely generated D_R-module. Namely, we have $\dim C(M) = d(M)$. In particular $C(M) = 0$ if and only if $M = 0$.

When our finitely generated D_R-module has a presentation $M = \frac{D_R}{L}$ we have a good filtration on M induced by the order filtration on D_R such that

$$gr_{(0,1)} M = \frac{k[x_1, \ldots, x_n, \xi_1, \ldots, \xi_n]}{gr_{(0,1)} L}.$$

Therefore, the characteristic variety is given by the ideal $J(M) = \mathrm{rad}\,(gr_{(0,1)}L)$, where $gr_{(0,1)}L$ is the ideal generated by the symbols $\sigma_{(0,1)}(P)$ for all $P \in L$.

- *Polynomial ring R:* Consider the presentation

$$R = \frac{D_R}{D_R(\partial_1, \ldots, \partial_n)}.$$

Therefore we have $J(R) = (\xi_1, \ldots, \xi_n)$.

- *Localizations at monomials:* Consider the presentation

$$R_{x_1 \cdots x_p} = \frac{D_R}{D_R(x_1\partial_1 + 1, \ldots, x_p\partial_p + 1, \partial_{p+1}, \ldots, \partial_n)}.$$

Therefore we have $J(R_{x_1 \cdots x_p}) = (x_1\xi_1, \ldots, x_p\xi_p, \xi_{p+1}, \ldots, \xi_n)$.

- *Local cohomology modules:* Consider the presentation

$$H^p_{(x_1, \ldots, x_p)}(R) = \frac{D_R}{D_R(x_1, \ldots, x_p, \partial_{p+1}, \ldots, \partial_n)}.$$

Therefore we have $J(H^p_{(x_1, \ldots, x_p)}(R)) = (x_1, \ldots, x_p, \xi_{p+1}, \ldots, \xi_n)$.

In particular, $J(H^n_{\mathfrak{m}}(R)) = (x_1, \ldots, x_n)$.

The characteristic variety of the polynomial ring R or the local cohomology modules $H^p_{(x_1, \ldots, x_p)}(R)$ are irreducible, but the characteristic variety of the localization $R_{x_1 \cdots x_p}$ is not. We can refine the characteristic variety with the so-called *characteristic cycle* that encodes its components with a certain multiplicity.

- *Characteristic cycle:* Is the formal sum

$$CC(M) = \sum m_i V_i$$

taken over all the irreducible components V_i of the characteristic variety $C(M)$ and the $m_i's$ are the multiplicities of $gr_\Gamma M$ at a generic point along each component V_i.

The multiplicities can also be described using the theory of Hilbert functions. Let $V_i = V(\mathfrak{p}_i) \subseteq C(M)$ be an irreducible component, where $\mathfrak{p}_i \in \mathrm{Spec}\,(gr_{(0,1)}D_R)$. Then m_i is the multiplicity of the module $gr_\Gamma M_{\mathfrak{p}_i}$. Namely, the Hilbert series:

$$H(gr_\Gamma M_{\mathfrak{p}_i}; t) = \sum_{j \geq 0} \dim_k [\mathfrak{p}_i^j gr_\Gamma M_{\mathfrak{p}_i} / \mathfrak{p}_i^{j+1} gr_\Gamma M_{\mathfrak{p}_i}]\, t^j$$

is in the form $H(gr_\Gamma M_{\mathfrak{p}_i}; t) = q_i(t)/(1 - t)^{d_i}$, where $q_i(t) \in \mathbb{Z}[t, t^{-1}]$ is such that $q_i(1) \neq 0$, so the multiplicity in the characteristic cycle of the irreducible component V_i is then $m_i = q_i(1)$.

Warning: If we consider the formal power series ring $S = k[[x_1, \ldots, x_n]]$ and its corresponding ring of differential operators D_S we can mimic all the above constructions. Thus we can define the class of holonomic D_S-modules and construct the corresponding characteristic cycle. The same can be done in the analytic case, i.e. when $S = \mathbb{C}\{x_1, \ldots, x_n\}$ is the ring of convergent series with complex coefficients.

However one must be careful with the components of the characteristic variety when we work over the polynomial ring $k[x_1, \ldots, x_n]$, the formal power series ring $k[[x_1, \ldots, x_n]]$ or the analytic case $\mathbb{C}\{x_1, \ldots, x_n\}$ since they may differ. In the analytic case, F. Pham [158] (see also [122]) completely described these components. His result states that the irreducible components are conormal bundles $T^*_{X_i} X$ relative to $X_i \subseteq X = \mathbb{C}^n$ so we have

$$CC(M) = \sum m_i T^*_{X_i} X.$$

The characteristic ideals of the examples we will use in these lectures are going to be monomial ideals so we will not have problems with their primary decomposition when viewed over the polynomial ring or over any series ring so there will be no problem borrowing the notation from the analytic case. In the sequel we will just denote

$$X = \operatorname{Spec} k[x_1, \ldots, x_n] = \mathbb{A}^n_k$$

$$T^*X = \operatorname{Spec} k[x_1, \ldots, x_n, \xi_1, \ldots, \xi_n] = \mathbb{A}^{2n}_k$$

$$T^*_{X_\alpha} X = V(\{x_i \mid \alpha_i = 1\}, \{\xi_i \mid \alpha_i = 0\}) \subseteq T^*X$$

where $X_\alpha = V(\mathfrak{p}_\alpha) \subseteq X$ is the variety defined by the homogeneous prime ideal $\mathfrak{p}_\alpha := (x_i \mid \alpha_i \neq 0)$, $\alpha \in \{0, 1\}^n$. We denote $T^*_X X = V(\xi_1, \ldots, \xi_n)$ for the case $\alpha = (0, \ldots, 0) \in \{0, 1\}^n$. This notation is very useful when we consider the projection $\pi : T^*X \longrightarrow X$ given by the map $\pi(\mathbf{x}, \boldsymbol{\xi}) = \mathbf{x}$, since

$$\pi(T^*_{X_i} X) = X_i.$$

1.7.1 Examples

In general, the multiplicities of the components of the characteristic variety might be difficult to compute but, when we have a presentation $M = \frac{D_R}{L}$ such that the ideal $gr_{(0,1)} L$ is radical, the associated multiplicities are 1. This is what happens with the examples we are dealing with.

- *Polynomial ring R:* The characteristic variety $C(R) = V(\xi_1, \ldots, \xi_n)$ has only a component and the associated multiplicity is 1. Therefore

$$CC(R) = T^*_X X.$$

- *Localizations at monomials:* The characteristic variety

$$C(R_{x_1 \cdots x_p}) = V(x_1 \xi_1, \ldots, x_p \xi_p, \xi_{p+1}, \ldots, \xi_n)$$

has 2^p components with associated multiplicity 1. Namely, we have

$$CC(R_{x_1}) = T_X^* X + T_{X_{(1,0,\ldots,0)}}^* X,$$

$$CC(R_{x_1 x_2}) = T_X^* X + T_{X_{(1,0,\ldots,0)}}^* X + T_{X_{(0,1,0,\ldots,0)}}^* X + T_{X_{(1,1,0,\ldots,0)}}^* X,$$

and in general

$$CC(R_{x_1 x_2 \cdots x_p}) = \sum_{\beta \leq \alpha} T_{X_\beta}^* X, \text{ where } \alpha = (\underbrace{1, \ldots, 1}_{p}, 0, \ldots, 0).$$

- *Local cohomology modules:* The characteristic variety

$$C(H_{(x_1,\ldots,x_p)}^p(R)) = V(x_1, \ldots, x_p, \xi_{p+1}, \ldots, \xi_n)$$

has only a component and the associated multiplicity is 1. Therefore

$$CC(H_{(x_1,\ldots,x_p)}^p(R)) = T_{X_\alpha}^* X, \text{ where } \alpha = (\underbrace{1, \ldots, 1}_{p}, 0, \ldots, 0).$$

In particular, $CC(H_{\mathfrak{m}}^n(R)) = T_{X_1}^* X$.

1.8 Applications

The characteristic cycle turns out to be a very useful tool in the study of D_R-modules. It is an invariant of the category of D_R-modules that also provides information on the object when viewed as R-module. Mainly, the varieties that appear in the formula $CC(M) = \sum m_\alpha T_{X_\alpha}^* X$ for any given holonomic D_R-module M describe the support of M as R-module, but we also get some extra information coming from the corresponding multiplicities.

- *Support as R-module:* Let

$$\pi : \text{Spec}\,(k[x_1, \ldots, x_n, \xi_1, \ldots, \xi_n]) \longrightarrow \text{Spec}\,(k[x_1, \ldots, x_n])$$

be the projection map defined by $\pi(x, \xi) = x$. Then, for any holonomic D_R-module, we have:

$$\text{Supp}_R(M) = \pi(C(M)).$$

Therefore, the notation that we use to describe the characteristic cycle will be very convenient. Namely, if $CC(M) = \sum m_\alpha T^*_{X_\alpha} X$ then $\text{Supp}_R(M) = \bigcup X_\alpha$.

In general, the characteristic cycle is difficult to compute directly. The following property will be very useful when computing the characteristic cycle of local cohomology modules via the Čech complex or the Mayer–Vietoris sequence.

- *Additivity of the characteristic cycle with respect to exact sequences:* Let $0 \longrightarrow M_1 \longrightarrow M_2 \longrightarrow M_3 \longrightarrow 0$ be an exact sequence of holonomic D_R-modules. Then, we have $CC(M_2) = CC(M_1) + CC(M_3)$.

When $R = k[[x_1, \dots, x_n]]$ we can describe Lyubeznik numbers using the characteristic cycle of the local cohomology module $H^p_\mathfrak{m}(H^{n-i}_I(R))$.

- *Lyubeznik numbers:* Recall that these invariants are defined as

$$\lambda_{p,i}(R/I) = \mu_p(\mathfrak{m}, H^{n-i}_I(R)) = \mu_0(\mathfrak{m}, H^p_\mathfrak{m}(H^{n-i}_I(R))),$$

so we have $H^p_\mathfrak{m}(H^{n-i}_I(R)) \cong E(R/\mathfrak{m})^{\lambda_{p,i}}$. From the isomorphism $E(R/\mathfrak{m}) \cong H^n_\mathfrak{m}(R)$ and the additivity of the characteristic cycle we get

$$CC(H^p_\mathfrak{m}(H^{n-i}_I(R))) = \lambda_{p,i} \ T^*_{X_1} X,$$

where X_1 is the variety defined by $\mathfrak{m} \subseteq R$.

- *Some invariants of local rings:* A generalization of Lyubeznik numbers has been given in [5] using the characteristic cycles of local cohomology modules. The proof of the following theorem is analogous to the proof of [132, Theorem 4.1] but one must be careful with the behavior of the characteristic cycle so some results on direct images of D_R-modules are required.

Theorem 4. *Let A be a ring which admits a surjective ring homomorphism $\pi :$ $R \longrightarrow A$, where R is regular local ring[4] of dimension n. Let $I = \ker \pi$ and $\mathfrak{p} \in$ Spec (R) such that $I \subseteq \mathfrak{p}$. Consider the characteristic cycles:*

- $CC(H^{n-i}_I(R)) = \sum m_{i,\alpha} \ T^*_{X_\alpha} X,$
- $CC(H^p_\mathfrak{p}(H^{n-i}_I(R))) = \sum \lambda_{\mathfrak{p},p,i,\alpha} \ T^*_{X_\alpha} X.$

Then, the following multiplicities do not depend neither on R nor on π:

- *The multiplicities $m_{i,\alpha}$ only depend on A, i and α.*
- *The multiplicities $\lambda_{\mathfrak{p},p,i,\alpha}$ only depend on A, $\mathfrak{p}$, p, i and α.*

Among these multiplicities we may find:

- *Bass numbers:* $\lambda_{\mathfrak{p},p,i,\alpha_\mathfrak{p}} = \mu_p(\mathfrak{p}, H^{n-i}_I(R))$, where $X_{\alpha_\mathfrak{p}}$ is the variety defined by $\mathfrak{p} \subseteq R$.
- *Lyubeznik numbers:* $\lambda_{\mathfrak{m},p,i,\alpha_\mathfrak{m}} = \lambda_{p,i}(A)$, where $X_{\alpha_\mathfrak{m}}$ is the variety defined by $\mathfrak{m} \subseteq R$.

---

[4]After completion we can always assume that $R = k[[x_1, \dots, x_n]]$ is the formal power series ring.

Collecting the multiplicities $m_{i,\alpha}$ of the characteristic cycle of $H_I^{n-i}(R)$ by the dimension of the corresponding varieties we get the coarser invariants:

$$\gamma_{p,i}(A) := \{\sum m_{i,\alpha} \mid \dim X_\alpha = p\}.$$

These invariants have the same properties as Lyubeznik numbers (see [5]). Namely, let $d = \dim A$. Then $\gamma_{d,d}(A) \neq 0$ and $\gamma_{p,i}(A) = 0$ if $i > d$, $p > i$ so we can also consider the following table

$$\Gamma(A) = \begin{pmatrix} \gamma_{0,0} & \cdots & \gamma_{0,d} \\ & \ddots & \vdots \\ & & \gamma_{d,d} \end{pmatrix}.$$

Notice that each column gives us information on the support of a local cohomology module $H_I^{n-i}(R)$, in particular we have

$$\dim_R \operatorname{Supp}_R H_I^{n-i}(R) = \max\{ p \mid \gamma_{p,i} \neq 0\}.$$

2 Local Cohomology Modules Supported on Monomial Ideals

Let $R = k[x_1, \ldots, x_n]$ be the polynomial ring in n independent variables, where k is a field. An ideal $I \subseteq R$ is said to be a squarefree monomial ideal if it may be generated by squarefree monomials $\mathbf{x}^\alpha := x_1^{\alpha_1} \cdots x_n^{\alpha_n}$, $\alpha \in \{0,1\}^n$. Its minimal primary decomposition is given in terms of face ideals $\mathfrak{p}_\alpha := (x_i \mid \alpha_i \neq 0)$, $\alpha \in \{0,1\}^n$. For simplicity we will denote the homogeneous maximal ideal $\mathfrak{m} := \mathfrak{p}_1 = (x_1, \ldots, x_n)$, where $\mathbf{1} = (1, \ldots, 1)$. As usual, we denote $|\alpha| = \alpha_1 + \cdots + \alpha_n$ and $\varepsilon_1, \ldots, \varepsilon_n$ will be the natural basis of $\mathbb{Z}^n$.

The Alexander dual ideal of I is the ideal $I^\vee = (\mathbf{x}^\alpha \mid \mathbf{x}^{1-\alpha} \notin I)$. The minimal primary decomposition of $I^\vee$ can be easily described from I. Namely, let $\{\mathbf{x}^{\alpha_1}, \ldots, \mathbf{x}^{\alpha_r}\}$ be a minimal system of generators of I. Then, the minimal primary decomposition of $I^\vee$ is of the form $I^\vee = \mathfrak{p}_{\alpha_1} \cap \cdots \cap \mathfrak{p}_{\alpha_r}$, and we have $I^{\vee\vee} = I$.

In this section we will study the structure of local cohomology modules $H_I^r(R)$ from different points of view. We will start studying its $\mathbb{Z}^n$-graded structure and then we will continue with the D_R-module structure. It turns out that, even they have a different nature, both approaches are equivalent. Finally we will see that the structure of the local cohomology modules $H_I^r(R)$ also allow us to describe the free resolution of the Alexander dual ideal $I^\vee$.

2.1 $\mathbb{Z}^n$-Graded Structure

The polynomial ring $R = k[x_1, \ldots, x_n]$ has a natural $\mathbb{Z}^n$-grading given by $\deg(x_i) = \varepsilon_i$. The quotients R/I, where $I \subseteq R$ is a monomial ideal, and the localizations at a squarefree monomial $R_{\mathbf{x}^\alpha}$, $\alpha \in \{0, 1\}^n$ inherit a natural $\mathbb{Z}^n$-graded structure. Then, by using the Čech complex, the local cohomology modules $H_{\mathfrak{m}}^r(R/I)$ and $H_I^r(R)$ also have a $\mathbb{Z}^n$-graded structure.

In the last decade or so there has been a lot of progress on the understanding of this $\mathbb{Z}^n$-graded structure but the germ of the theory is the fundamental theorem of M. Hochster, that finally appeared in [182, Theorem II 4.1], where he gives a description of the Hilbert series of $H_{\mathfrak{m}}^r(R/I)$.

To describe this formula we will make use of the *Stanley–Reisner correspondence* that states that to any squarefree monomial ideal $I \subseteq R$ one associates a simplicial complex Δ defined over the set of vertices $\{x_1, \ldots, x_n\}$ such that $I = I_\Delta := (\mathbf{x}^\alpha \mid \sigma_\alpha \notin \Delta)$, where $\sigma_\alpha := \{x_i \mid \alpha_i = 1\}$ for $\alpha \in \{0, 1\}^n$. We point out that this correspondence is compatible with Alexander duality in the sense that $I_\Delta^\vee = I_{\Delta^\vee}$, where the Alexander dual simplicial complex is $\Delta^\vee := \{\sigma_{1-\alpha} \mid \sigma_\alpha \notin \Delta\}$, i.e. $\Delta^\vee$ consists of the complements of the nonfaces of Δ.

In this section are also going to use the following subcomplexes associated to the face $\sigma_\alpha \in \Delta$, $\alpha \in \{0, 1\}^n$:

- *Restriction to σ_α:* $\Delta_\alpha := \{\tau \in \Delta \mid \tau \in \sigma_\alpha\}$.
- *Link of σ_α:* $\operatorname{link}_\alpha \Delta := \{\tau \in \Delta \mid \sigma_\alpha \cap \tau = \emptyset, \ \sigma_\alpha \cup \tau \in \Delta\}$.

We have to point out that the equality of complexes $\Delta_{1-\alpha}^\vee = (\operatorname{link}_\alpha \Delta)^\vee$ and Alexander duality provide an isomorphism of reduced simplicial (co-)homology groups:

$$\tilde{H}_{n-|\alpha|-r-1}(\operatorname{link}_\alpha \Delta; k) \cong \tilde{H}^{r-2}(\Delta_{1-\alpha}^\vee; k).$$

2.2 $\mathbb{Z}^n$-Graded Structure of $H_{\mathfrak{m}}^r(R/I)$

M. Hochster's formula for the $\mathbb{Z}^n$-graded Hilbert series of the local cohomology modules $H_{\mathfrak{m}}^r(R/I)$ is expressed in terms of the reduced simplicial cohomology of links of the simplicial complex Δ associated to the ideal I.

Theorem 5 (Hochster). *Let $I = I_\Delta$ be the Stanley–Reisner ideal of a simplicial complex Δ. Then, the $\mathbb{Z}^n$-graded Hilbert series of $H_{\mathfrak{m}}^r(R/I)$ is:*

$$H(H_{\mathfrak{m}}^r(R/I); \mathbf{x}) = \sum_{\sigma_\alpha \in \Delta} \dim_k \tilde{H}^{r-|\alpha|-1}(\operatorname{link}_\alpha \Delta; k) \prod_{\alpha_i = 1} \frac{x_i^{-1}}{1 - x_i^{-1}}.$$

From M. Hochster's formula we deduce the isomorphisms

$$H_{\mathfrak{m}}^r(R/I)_\beta \cong \tilde{H}^{r-|\alpha|-1}(\operatorname{link}_\alpha \Delta; k), \quad \forall \beta \in \mathbb{Z}^n \text{ such that } \sigma_\alpha = \sup_-(\beta),$$

where $\sup_-(\beta) := \{x_i \mid \beta_i < 0\}$. We also deduce that the multiplication by the variable x_i establishes an isomorphism between the pieces $H_{\mathfrak{m}}^r(R/I)_\beta$ and $H_{\mathfrak{m}}^r(R/I)_{\beta+\varepsilon_i}$ for all $\beta \in \mathbb{Z}^n$ such that $\beta_i \neq -1$.

Notice then that, in order to describe the $\mathbb{Z}^n$-graded structure of this module, we only have to determine the multiplication by x_i on the pieces $H_{\mathfrak{m}}^r(R/I)_{-\alpha}$, $\alpha \in \{0,1\}^n$. H. G. Gräbe [79], gave a topological interpretation of these multiplications.

Theorem 6 (Gräbe). *For all $\alpha \in \{0,1\}^n$ such that $\sigma_\alpha \in \Delta$, the morphism of multiplication by the variable x_i:*

$$\cdot x_i : H_{\mathfrak{m}}^r(R/I)_{-\alpha} \longrightarrow H_{\mathfrak{m}}^r(R/I)_{-(\alpha-\varepsilon_i)}$$

corresponds to the morphism

$$\tilde{H}^{r-|\alpha|-1}(\operatorname{link}_\alpha \Delta; k) \longrightarrow \tilde{H}^{r-|\alpha-\varepsilon_i|-1}(\operatorname{link}_{\alpha-\varepsilon_i} \Delta; k),$$

or equivalently the morphism

$$\tilde{H}^{r-2}(\Delta_{1-\alpha}^\vee; k) \longrightarrow \tilde{H}^{r-2}(\Delta_{1-\alpha+\varepsilon_i}^\vee; k),$$

induced by the inclusion $\Delta_{1-\alpha+\varepsilon_i}^\vee \subseteq \Delta_{1-\alpha}^\vee$.

2.3 $\mathbb{Z}^n$-Graded Structure of $H_I^r(R)$

Inspired by M. Hochster's formula, N. Terai [187] gave a description of the $\mathbb{Z}^n$-graded Hilbert series of the local cohomology modules $H_I^r(R)$, in this case expressed in terms of the reduced simplicial homology of the links.

Theorem 7 (Terai). *Let $I = I_\Delta$ be the Stanley–Reisner ideal of a simplicial complex Δ. Then, the graded Hilbert series of $H_I^r(R)$ is:*

$$H(H_I^r(R); \mathbf{x}) = \sum_{\alpha \in \{0,1\}^n} \dim_k \tilde{H}_{n-r-|\alpha|-1}(\operatorname{link}_\alpha \Delta; k) \prod_{\alpha_i=0} \frac{x_i^{-1}}{1-x_i^{-1}} \prod_{\alpha_j=1} \frac{1}{1-x_j}.$$

From N. Terai's formula one also may deduce the isomorphisms

$$H_I^r(R)_\beta \cong \tilde{H}_{n-r-|\alpha|-1}(\operatorname{link}_\alpha \Delta; k), \quad \forall \beta \in \mathbb{Z}^n \text{ such that } \sigma_\alpha = \sup_-(\beta)$$

and that the multiplication by the variable x_i establishes an isomorphism between the graded pieces $H_I^r(R)_\beta$ and $H_I^r(R)_{\beta+\varepsilon_i}$ for all $\beta \in \mathbb{Z}^n$ such that $\beta_i \neq -1$.

At the same time and independently, M. Mustaţă [146] also described the pieces of the local cohomology modules $H_I^r(R)$ but he also gave a topological interpretation of the multiplication by x_i on the pieces $H_I^r(R)_{-\alpha}$, $\alpha \in \{0, 1\}^n$.

Theorem 8 (Mustaţă). *Let* $I = I_\Delta$ *be the Stanley–Reisner ideal of a simplicial complex* Δ. *Then,*

$$H_I^r(R)_\beta \cong \tilde{H}^{r-2}(\Delta_{1-\alpha}^\vee; k), \quad \forall \beta \in \mathbb{Z}^n \text{ such that } \sigma_\alpha = \sup_-(\beta).$$

Moreover, for all $\alpha \in \{0, 1\}^n$ *such that* $\sigma_\alpha \in \Delta$, *the morphism of multiplication by the variable* x_i:

$$\cdot x_i : H_I^r(R)_{-\alpha} \longrightarrow H_I^r(R)_{-(\alpha - \varepsilon_i)}$$

corresponds to the morphism

$$\tilde{H}^{r-2}(\Delta_{1-\alpha}^\vee; k) \longrightarrow \tilde{H}^{r-2}(\Delta_{1-\alpha+\varepsilon_i}^\vee; k),$$

induced by the inclusion $\Delta_{1-\alpha+\varepsilon_i}^\vee \subseteq \Delta_{1-\alpha}^\vee$.

We remark that the formulas of M. Hochster and N. Terai are equivalent by using the Čech hull and Alexander duality (see [140]). The same happens with the formulas of H. G. Gräbe and M. Mustaţă.

2.4 A General Framework: Squarefree and Straight Modules

K. Yanagawa [199] introduced the notion of squarefree modules over a polynomial ring $R = k[x_1, \ldots, x_n]$ to generalize the theory of Stanley–Reisner rings. In this setting one can apply homological methods to study monomial ideals in a more systematical way. We recall his definition.

Definition 9 ([199]). A $\mathbb{N}^n$-graded module M is said to be **squarefree** if the following two conditions are satisfied:

i) $\dim_k M_\alpha < \infty$ for all $\alpha \in \mathbb{Z}^n$.
ii) The multiplication map $M_\alpha \ni y \mapsto \mathbf{x}^\beta y \in M_{\alpha+\beta}$ is bijective for all $\alpha, \beta \in \mathbb{N}^n$ with $\text{supp}(\alpha + \beta) = \text{supp}(\alpha)$.

A squarefree monomial ideal I and the corresponding quotient ring R/I are squarefree modules. A free module $R(-\alpha)$ shifted by $\alpha \in \{0, 1\}^n$, is also squarefree. In particular, the $\mathbb{Z}^n$-graded canonical module $\omega_R = R(-1)$ of R is squarefree, where $\mathbf{1} = (1, \ldots, 1)$. To describe the $\mathbb{N}^n$-graded structure of a squarefree module M, one only needs to describe the pieces M_α, $\alpha \in \{0, 1\}^n$ and the multiplication maps $x_i : M_\alpha \longrightarrow M_{\alpha+\varepsilon_i}$.

The full subcategory of the category *Mod(R) of $\mathbb{Z}^n$-graded R-modules which has as objects the squarefree modules will be denoted **Sq**. This is an abelian category closed by kernels, cokernels and extensions. It has enough injectives and projectives modules so one can develop all the usual tools in homological algebra. A more precise description of these objects is as follows:

- *Simple:* $\operatorname{Ext}_R^{|\alpha|}(R/\mathfrak{p}_\alpha, \omega_R)$ for any face ideal $\mathfrak{p}_\alpha$, $\alpha \in \{0, 1\}^n$.
- *Injective:* $R/\mathfrak{p}_\alpha$ for any face ideal $\mathfrak{p}_\alpha$, $\alpha \in \{0, 1\}^n$.
- *Projective:* $R(-\alpha)$, $\alpha \in \{0, 1\}^n$.

Building on the previous concept, K. Yanagawa [200] also developed a similar notion for $\mathbb{Z}^n$-graded modules.

Definition 10 ([200]). A $\mathbb{Z}^n$-graded module M is said to be **straight** if the following two conditions are satisfied:

i) $\dim_k M_\alpha < \infty$ for all $\alpha \in \mathbb{Z}^n$.
ii) The multiplication map $M_\alpha \ni y \mapsto \mathbf{x}^\beta y \in M_{\alpha+\beta}$ is bijective for all $\alpha, \beta \in \mathbb{Z}^n$ with $\operatorname{supp}(\alpha + \beta) = \operatorname{supp}(\alpha)$.

The main example of straight modules are the local cohomology modules of the canonical module $H_I^r(\omega_R)$ supported on a monomial ideal $I \subseteq R$. Again, in order to describe the $\mathbb{Z}^n$-graded structure of a straight module M, one has to describe the pieces M_α, $\alpha \in \{0, 1\}^n$ and the multiplication maps $x_i : M_\alpha \longrightarrow M_{\alpha+\varepsilon_i}$.

The full subcategory of the category $^*\mathrm{Mod}(R)$ of $\mathbb{Z}^n$-graded R-modules which has as objects the straight modules will be denoted **Str**. This is an abelian category closed by kernels, cokernels and extensions with enough injectives and projectives described as follows:

- *Simple:* $H_{\mathfrak{p}_\alpha}^{|\alpha|}(\omega_R)$ for any face ideal $\mathfrak{p}_\alpha$, $\alpha \in \{0, 1\}^n$.
- *Injective:* $^*E(R/\mathfrak{p}_\alpha)$ for any face ideal $\mathfrak{p}_\alpha$, $\alpha \in \{0, 1\}^n$.
- *Projective:* $R_{\mathbf{x}^\alpha}(-\mathbf{1})$, $\alpha \in \{0, 1\}^n$.

A slight variation of [200, Proposition 2.12] gives a nice characterization of these modules in terms of the following filtration

Proposition 11. *A $\mathbb{Z}^n$-graded module M is straight if and only if there is an increasing filtration $\{F_j\}_{0 \le j \le n}$ of M by $\mathbb{Z}^n$-graded submodules and there are integers $m_\alpha \ge 0$ for $\alpha \in \{0, 1\}^n$, such that for all $0 \le j \le n$ one has isomorphisms*

$$F_j/F_{j-1} \cong \bigoplus_{\substack{\alpha \in \{0,1\}^n \\ |\alpha|=j}} (H_{\mathfrak{p}_\alpha}^j(\omega_R))^{\oplus m_\alpha}.$$

Therefore we obtain a set of short exact sequences

$$
\begin{aligned}
(s_1): & \quad 0 \to F_0 \to F_1 \to F_1/F_0 \to 0 \\
(s_2): & \quad 0 \to F_1 \to F_2 \to F_2/F_1 \to 0 \\
& \qquad\qquad \vdots \\
(s_n): & \quad 0 \to F_{n-1} \to M \to F_n/F_{n-1} \to 0
\end{aligned}
$$

The extension classes of these exact sequences determine the structure of the straight module. Each extension class of the sequence (s_j) defines an element in $^*\mathrm{Ext}^1_R(F_j/F_{j-1}, F_{j-1})$.

For a $\mathbb{Z}^n$-graded R-module $M = \bigoplus_{\alpha \in \mathbb{Z}^n} M_\alpha$, we call the submodule $\bigoplus_{\alpha \in \mathbb{N}^n} M_\alpha$ the $\mathbb{N}^n$-graded part of M, and denote it by $\mathscr{N}(M)$. If M is straight then $\mathscr{N}(M)$ is squarefree. Conversely, for any squarefree module N, there is a unique (up to isomorphism) straight module $\mathscr{Z}(N)$ whose $\mathbb{N}^n$-graded part is isomorphic to N. For example,

$$\mathscr{Z}(\mathrm{Ext}^{|\alpha|}_R(R/\mathfrak{p}_\alpha, \omega_R)) = H^{|\alpha|}_{\mathfrak{p}_\alpha}(\omega_R).$$

It was proved in [200, Proposition 2.7] that the functors $\mathscr{N} : \mathbf{Str} \longrightarrow \mathbf{Sq}$ and $\mathscr{Z} : \mathbf{Sq} \longrightarrow \mathbf{Str}$ establish an equivalence of categories between squarefree and straight modules. For further considerations and generalizations of this theory we recommend to take a look at [201, 202].

A generalization of squarefree and straight modules was given by E. Miller [140] (see also [141]). In his terminology positively 1-determined modules correspond to squarefree modules and 1-determined module corresponds to straight modules. In this generalized framework he introduced the *Alexander duality functors* that are closely related to Matlis duality and local duality. For the case of squarefree modules and independently, T. Römer [166] also introduced Alexander duality via the exterior algebra.

2.5 D-Module Structure

Recently, there has been made an effort towards effective computation of local cohomology modules by using the theory of Gröbner bases over rings of differential operators. Algorithms given by U. Walther [194] and T. Oaku and N. Takayama [153] provide a utility for such computation and are both implemented in the package D-modules [127] for Macaulay 2 [80].

U. Walther's algorithm is based on the construction of the Čech complex of holonomic D_R-modules. So it is necessary to give a description of the localization R_f at a polynomial $f \in R$. An algorithm to compute these modules was given by T. Oaku in [152]. The main ingredient of the algorithm is the computation of the Bernstein–Sato polynomial of f which turns out to be a major bottleneck due to its complexity. For some short examples we can do the job just using the Macaulay2 command localCohom.

The first goal of this section is to compute the characteristic cycle of local cohomology modules supported on monomial ideals. Our aim is to avoid a direct computation using the additivity with respect to exact sequences. Recall that, to compute this invariant directly one needs to:

- Construct a presentation of the D_R-module $H_I^r(R)$,
- Compute the characteristic ideal $J(H_I^r(R))$,
- Compute the primary components of $J(H_I^r(R))$ and its multiplicities.

As we said, we can work out the first step using the Macaulay2 command localCohom for some short examples and we can also use the command charIdeal to compute its characteristic ideal.

2.6 Characteristic Cycle of $H_I^r(R)$

Let $R = k[x_1, \ldots, x_n]$ be the polynomial ring over a field k of characteristic zero. The aim of this section is to compute the characteristic cycle of the local cohomology modules $H_I^r(R)$ for any given squarefree monomial ideal $I \subseteq R$.

We want to use the additivity of the characteristic cycle with respect to short exact sequences to reduce the problem to the computation of the characteristic cycles of some building blocks. These building blocks are going to be localizations of the polynomial ring at monomials when using the Čech complex. On the other hand, if we use the Mayer–Vietoris sequence, the building blocks are going to be local cohomology modules supported on face ideals. In any case, the characteristic cycle of these holonomic D_R-modules have already been computed in Sect. 1.6.1.

2.6.1 Using the Čech Complex

Let $I = (f_1, \ldots, f_s) \subseteq R$, be a squarefree monomial ideal, i.e. the generators f_i are monomials of the form $\mathbf{x}^\beta$, $\beta \in \{0, 1\}^n$. Consider the Čech complex

$$\check{C}_I^\bullet(R): \quad 0 \longrightarrow R \xrightarrow{d_0} \oplus_{1 \le i \le s} R_{f_i} \xrightarrow{d_1} \cdots \xrightarrow{d_{s-1}} R_{f_1 \cdots f_s} \longrightarrow 0$$

where the differentials d_p are defined by using the canonical localization morphism on every component. This is a complex of flat $\mathbb{Z}^n$-graded modules so the differentials can be described using the so-called *monomial matrices* introduced by E. Miller [140].

Example 12. Let $I = (f_1, f_2, f_3) \subseteq R$, be a squarefree monomial ideal. Then, the Čech complex can be described as

$$
0 \longrightarrow R \xrightarrow{\begin{pmatrix} 1 \\ 1 \\ 1 \end{pmatrix}}
\begin{matrix} R_{f_1} \\ \oplus \\ R_{f_2} \\ \oplus \\ R_{f_3} \end{matrix}
\xrightarrow{\begin{pmatrix} -1 & 1 & 0 \\ -1 & 0 & 1 \\ 0 & -1 & 1 \end{pmatrix}}
\begin{matrix} R_{f_1 f_2} \\ \oplus \\ R_{f_1 f_3} \\ \oplus \\ R_{f_2 f_3} \end{matrix}
\xrightarrow{(-1,1,-1)}
R_{f_1 f_2 f_3} \longrightarrow 0
$$

The source and the target of these monomial matrices are labelled by the monomials f_1, f_2, f_3 and their products.

Our aim is to use these monomial matrices to construct a complex of k-vector spaces associated to each possible component $T^*_{X_\alpha} X$, $\alpha \in \{0, 1\}^n$ of the characteristic cycle such that the corresponding homology groups describe the corresponding multiplicity. To illustrate these computations we present the following:

Example 13. Consider the ideal $I = (x_1 x_2, x_1 x_3, x_2 x_3)$ in $R = k[x_1, x_2, x_3]$. We have the Čech complex $\check{C}^\bullet_I(R)$:

$$
0 \longrightarrow R \xrightarrow{\begin{pmatrix} 1 \\ 1 \\ 1 \end{pmatrix}} \begin{matrix} R_{x_1 x_2} \\ \oplus \\ R_{x_1 x_3} \\ \oplus \\ R_{x_2 x_3} \end{matrix} \xrightarrow{\begin{pmatrix} -1 & 1 & 0 \\ -1 & 0 & 1 \\ 0 & -1 & 1 \end{pmatrix}} \begin{matrix} R_{x_1 x_2 x_3} \\ \oplus \\ R_{x_1 x_2 x_3} \\ \oplus \\ R_{x_1 x_2 x_3} \end{matrix} \xrightarrow{(-1,1,-1)} R_{x_1 x_2 x_3} \longrightarrow 0
$$

The interested reader should try to figure out the labels for the source and target of the monomial matrices following the convention given in [140].

To compute the characteristic cycle of the local cohomology modules $H^r_I(R)$ we have to split the Čech complex into short exact sequences and use the additivity property but we can do all in once just keeping track of any component $T^*_{X_\alpha} X$, $\alpha \in \{0, 1\}^3$ that appear in the characteristic cycle of the localizations in the Čech complex.

- For $\alpha = (0, 0, 0)$ the component $T^*_X X$ appears in the characteristic cycle of every localization. We illustrate this fact in the following diagram:

$$
0 \longrightarrow T^*_X X \longrightarrow \begin{matrix} T^*_X X \\ + \\ T^*_X X \\ + \\ T^*_X X \end{matrix} \longrightarrow \begin{matrix} T^*_X X \\ + \\ T^*_X X \\ + \\ T^*_X X \end{matrix} \longrightarrow T^*_X X \longrightarrow 0
$$

One can check out that the multiplicity of $T^*_X X$ in the characteristic cycle of $H^r_I(R)$ is the dimension of the r-th cohomology groups of the following complex of k-vector spaces that we can construct using the monomial matrices describing the Čech complex.[5]

---

[5]One has to interpret the non-zero entries in the matrix as inclusions of the corresponding components of the Čech complex.

$$
0 \longrightarrow k \xrightarrow{\begin{pmatrix}1\\1\\1\end{pmatrix}} \begin{array}{c} k \\ \oplus \\ k \\ \oplus \\ k \end{array} \xrightarrow{\begin{pmatrix}-1 & 1 & 0\\-1 & 0 & 1\\0 & -1 & 1\end{pmatrix}} \begin{array}{c} k \\ \oplus \\ k \\ \oplus \\ k \end{array} \xrightarrow{(-1,1,-1)} k \longrightarrow 0
$$

This complex is acyclic so $T_X^* X$ is not a summand of the characteristic cycle of $H_I^r(R)$ for any r.

- For $\alpha = (1, 1, 0)$ the component $T_{X_{(1,1,0)}}^* X$ shows up in the following places:

$$
0 \longrightarrow 0 \longrightarrow \begin{array}{c} T_{X_{(1,1,0)}}^* X \\ + \\ 0 \\ + \\ 0 \end{array} \longrightarrow \begin{array}{c} T_{X_{(1,1,0)}}^* X \\ + \\ T_{X_{(1,1,0)}}^* X \\ + \\ T_{X_{(1,1,0)}}^* X \end{array} \longrightarrow T_{X_{(1,1,0)}}^* X \longrightarrow 0
$$

The complex of k-vector spaces that we obtain in this case is

$$
0 \longrightarrow 0 \longrightarrow k \xrightarrow{\begin{pmatrix}-1\\-1\\0\end{pmatrix}} k^3 \xrightarrow{(-1,1,-1)} k \longrightarrow 0
$$

so the characteristic cycle $CC(H_I^2(R))$ contains $T_{X_{(1,1,0)}}^* X$ with multiplicity 1. The cases $\alpha = (0, 1, 1), (1, 0, 1)$ are analogous.

- For $\alpha = (1, 1, 1)$ the component $T_{X_{(1,1,1)}}^* X$ appears in the following places:

$$
0 \longrightarrow 0 \longrightarrow \begin{array}{c} 0 \\ + \\ 0 \\ + \\ 0 \end{array} \longrightarrow \begin{array}{c} T_{X_{(1,1,1)}}^* X \\ + \\ T_{X_{(1,1,1)}}^* X \\ + \\ T_{X_{(1,1,1)}}^* X \end{array} \longrightarrow T_{X_{(1,1,1)}}^* X \longrightarrow 0
$$

The complex of k-vector spaces that we obtain in this case is

$$
0 \longrightarrow 0 \longrightarrow 0 \longrightarrow k^3 \xrightarrow{(-1,1,-1)} k \longrightarrow 0
$$

so $CC(H_I^2(R))$ contains $T_{X_{(1,1,1)}}^* X$ with multiplicity 2.

Therefore there is a local cohomology module different from zero and its characteristic cycle is

$$CC(H_I^2(R)) = T_{X_{(1,1,0)}}^* X + T_{X_{(1,0,1)}}^* X + T_{X_{(0,1,1)}}^* X + 2\, T_{X_{(1,1,1)}}^* X.$$

General case: Let $I = (f_1, \ldots, f_s) \subseteq R$ be a squarefree monomial ideal, and the generators f_i are monomials of the form $\mathbf{x}^\beta$, $\beta \in \{0,1\}^n$. Consider the Čech complex

$$\check{C}_I^\bullet(R): \quad 0 \longrightarrow R \xrightarrow{d_0} \oplus_{1 \le i \le s} R_{f_i} \xrightarrow{d_1} \cdots \xrightarrow{d_{s-1}} R_{f_1 \cdots f_s} \longrightarrow 0.$$

with the corresponding monomial matrices that describe the morphisms in the complex. Then, for any $\alpha \in \{0,1\}^n$ we construct a complex of k-vector spaces $[\check{C}_I^\bullet(R)]_\alpha$ that encodes when the component $T_{X_\alpha}^* X$ appears in the characteristic cycle of the localizations in the Čech complex, i.e. we have a copy of the field k in each position where the component $T_{X_\alpha}^* X$ appear. The morphisms of this complex are given by the monomial matrices that describe the Čech complex. Recall that the characteristic cycle of the localization of R at a monomial $\mathbf{x}^\beta$ is

$$CC(R_{\mathbf{x}^\beta}) = \sum_{\alpha \le \beta} T_{X_\alpha}^* X.$$

Then:

- For $\alpha = \mathbf{0} = (0, \ldots, 0) \in \{0,1\}^n$, since every localization $R_{\mathbf{x}^\beta}$ contains the component $T_X^* X$, the complex of k-vector spaces associated to the Čech complex $\check{C}_I^\bullet$ is:

$$[\check{C}_I^\bullet(R)]_0: \quad 0 \longrightarrow k \xrightarrow{d_0} k^s \xrightarrow{d_1} \cdots \xrightarrow{d_{s-1}} k \longrightarrow 0 .$$

This complex may be identified with the augmented relative simplicial cochain complex $\tilde{\mathscr{C}}^\bullet(\Delta_s; k)$, where Δ_s is the full simplicial complex whose vertices $\{x_1, \ldots, x_s\}$ are labeled by the minimal system of generators of I.

- In general, for any $\alpha \in \{0,1\}^n$, the component $T_{X_\alpha}^* X$ only appears in the localizations $R_{\mathbf{x}^\beta}$ such that $\beta \ge \alpha$ so the complex we construct is a subcomplex of $[\check{C}_I^\bullet(R)]_0$ and the morphisms are the corresponding restrictions. In order to give a topological interpretation, notice that, from the augmented relative simplicial chain complex $\tilde{\mathscr{C}}_\bullet(\Delta_s; k)$, we are taking out the pieces corresponding to the faces

$$\sigma_{1-\beta} := \{x_1, \ldots, x_s\} \setminus \{x_i \mid \beta_i = 1\} \in \Delta_s \quad \text{such that } \beta \not\ge \alpha.$$

Let $T_\alpha := \{\sigma_{1-\beta} \in \Delta_s \mid \beta \not\ge \alpha\}$ be a simplicial subcomplex of Δ_s. Then, the complex $[\check{C}_I^\bullet(R)]_\alpha$ may be identified with the augmented relative simplicial chain complex $\widetilde{\mathscr{C}}_\bullet(\Delta_s, T_\alpha; k)$ associated to the pair (Δ_s, T_α). By taking homology, the multiplicity of the component $T^*_{X_\alpha} X$ in the characteristic cycle of the local cohomology modules $H^r_I(R)$ are:

$$m_{n-r,\alpha} = \dim \tilde{H}_{r-1}(\Delta_s, T_\alpha; k) = \dim \tilde{H}_{r-2}(T_\alpha; k),$$

where the last assertion comes from the fact that Δ_s is contractible.

Proposition 14. *Let $I = (f_1, \ldots, f_s) \subseteq R$ be a squarefree monomial ideal. Then, the characteristic cycle of the local cohomology modules $H^r_I(R)$ is*

$$CC(H^r_I(R)) = \sum m_{n-r,\alpha} T^*_{X_\alpha} X$$

where $m_{n-r,\alpha} = \dim \tilde{H}_{r-2}(T_\alpha; k)$ and $T_\alpha := \{\sigma_{1-\beta} \in \Delta_s \mid \beta \not\geq \alpha\} \subseteq \Delta_s$.

Remark 15. The proof of this proposition follows from the additivity of the characteristic cycle and the fact that the monomial matrices that we use to construct the complex of k-vector spaces describe the inclusions between the localizations in the Čech complex.

Using the techniques we will develop in Sect. 2.8 one has a more direct proof since we have an isomorphism of complexes of k-vector spaces $[\check{C}^\bullet_I(R)]_\alpha = \mathrm{Hom}_{D_R}(\check{C}^\bullet_I(R), E_\alpha)$. We recommend the reader to go back to this point after getting familiar with the theory of n-hypercubes.

2.6.2 Using the Mayer–Vietoris Sequence

The usual method to compute local cohomology modules $H^r_I(R)$ is to find a representation of the ideal $I = U \cap V$ as the intersection of two simpler ideals U and V and then apply the Mayer–Vietoris sequence

$$\cdots \longrightarrow H^r_{U+V}(R) \longrightarrow H^r_U(R) \oplus H^r_V(R) \longrightarrow H^r_{U\cap V}(R) \longrightarrow H^{r+1}_{U+V}(R) \longrightarrow \cdots$$

In general, there are several choices for such a representation but in the case of squarefree monomial ideals we can use the good properties of the minimal primary decomposition $I = I_1 \cap \cdots \cap I_m$ and develop a method that will allow us to study these local cohomology modules in a systematical way. This is the approach used in [3] but we have to point out that we do not obtain a closed formula for the characteristic cycle of the local cohomology modules as the one we obtained in the previous subsection. The formula obtained in [3] comes after applying an algorithm that describes this iterated Mayer–Vietoris process.

We illustrate the method with the same example we used before.

Example 16. Consider the minimal primary decomposition

$$I = I_1 \cap I_2 \cap I_3 = (x_1, x_2) \cap (x_1, x_3) \cap (x_2, x_3)$$

of the squarefree monomial ideal $I = (x_1x_2, x_1x_3, x_2x_3)$ in $R = k[x_1, x_2, x_3]$. To study the local cohomology modules $H_I^r(R)$ we first use a Mayer–Vietoris sequence with:

$$U = I_1 \cap I_2, \qquad U \cap V = I = I_1 \cap I_2 \cap I_3,$$
$$V = I_3, \qquad\qquad U + V = (I_1 \cap I_2) + I_3.$$

We get the long exact sequence:

$$\cdots \longrightarrow H_{I_1 \cap I_2}^r(R) \oplus H_{I_3}^r(R) \longrightarrow H_I^r(R) \longrightarrow H_{(I_1 \cap I_2) + I_3}^{r+1}(R) \longrightarrow \cdots$$

- The ideal $I_1 \cap I_2 = (x_1, x_2x_3)$ is not a face ideal but we can describe the modules $H_{I_1 \cap I_2}^r(R)$ by using a Mayer–Vietoris sequence with:
$$U = I_1, \qquad U \cap V = I_1 \cap I_2,$$
$$V = I_2, \qquad U + V = I_1 + I_2.$$
- In general, not for this example, the ideal $(I_1 \cap I_2) + I_3$ is not a face ideal but we can describe the modules $H_{(I_1 \cap I_2) + I_3}^r(R)$ by using a Mayer–Vietoris sequence with:
$$U = I_1 + I_3, \qquad U \cap V = (I_1 \cap I_2) + I_3,$$
$$V = I_2 + I_3, \qquad U + V = I_1 + I_2 + I_3.$$

We can reflect the above process in the following diagram:

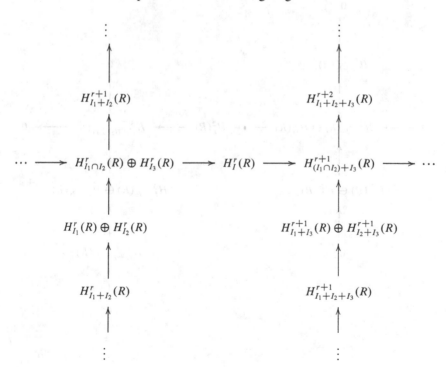

Thus, in order to describe the local cohomology modules $H_I^r(R)$, we have to study the modules:

$$H_{I_1}^r(R), \qquad H_{I_1+I_2}^r(R), \qquad H_{I_1+I_2+I_3}^r(R),$$
$$H_{I_2}^r(R), \qquad H_{I_1+I_3}^r(R),$$
$$H_{I_3}^r(R), \qquad H_{I_2+I_3}^r(R),$$

and the homomorphisms of the corresponding Mayer–Vietoris sequences. These modules are the local cohomology modules supported on all the ideals we can construct as sums of face ideals in the minimal primary decomposition of I. We state that these are the *initial pieces* that allow us to describe the modules $H_I^r(R)$. These sums of face ideals are again face ideals so they only have a non-vanishing local cohomology module. In our example, we have

$$I_1 + I_2 = I_1 + I_3 = I_2 + I_3 = I_1 + I_2 + I_3 = (x_1, x_2, x_3),$$

so the non-vanishing modules in this Mayer–Vietoris process are

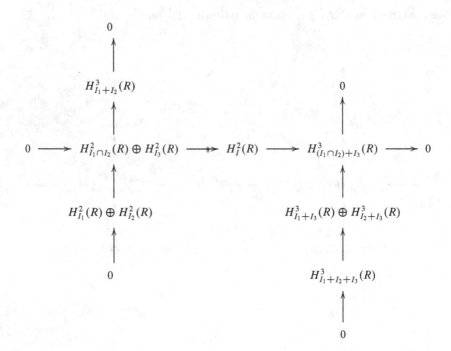

By the additivity of the characteristic cycle we get:

$$CC(H_I^2(R)) = CC(H_{(x_1,x_2)}^2(R)) + CC(H_{(x_1,x_3)}^2(R)) + CC(H_{(x_2,x_3)}^2(R))$$

$$+2 \; CC(H_{(x_1,x_2,x_3)}^3(R))$$

$$= T_{X_{(1,1,0)}}^* X + T_{X_{(1,0,1)}}^* X + T_{X_{(0,1,1)}}^* X + 2T_{X_{(1,1,1)}}^* X.$$

The general case is developed in [3] where an algorithm to get the characteristic cycle is given. The process is a lot more involved so we will skip the details. We will only point out that, given the minimal primary decomposition $I = I_1 \cap \cdots \cap I_m$, the initial pieces that describe the Mayer–Vietoris process are local cohomology modules supported on all the sums of face ideals in the minimal primary decomposition. The key point in the whole process is to control the kernels and cokernels that appear when we split all the Mayer–Vietoris sequences into short exact sequences.

2.6.3 Using the Mayer–Vietoris Spectral Sequence

All the Mayer–Vietoris process described above can be done all in once with the help of a spectral sequence introduced in [7] and developed in [135]. We do not assume the reader to have experience with the use of spectral sequence so we will skip the construction of this one and the meaning of *degeneration at the E_2-term* that leads to a closed formula for the characteristic cycle of local cohomology modules. For those that want to get more insight on this useful tool we recommend to take a look at a good homological algebra book for the basics on spectral sequences and then go to the details for this particular case in [7] or [135].

Let $I = I_1 \cap \cdots \cap I_m$ be a minimal primary decomposition of a squarefree monomial ideal $I \subseteq R$. Then we have the spectral sequence:

$$E_1^{-i,j} = \bigoplus_{1 \le \ell_1 < \cdots < \ell_i \le m} H_{I_{\ell_1} + \cdots + I_{\ell_i}}^j (R) \Longrightarrow H_I^{j-i}(R).$$

The E_1-page encodes the information given by the initial pieces we considered in the previous section. The E_2-page is more sophisticated and we need to introduce some notation.

The ideal I can be thought as the defining ideal of an arrangement $\mathscr{A}$ of linear varieties.[6] It defines a poset, i.e. a partial ordered set, $P(\mathscr{A})$ formed by the intersections of the irreducible components of X and the order given by the inclusion, i.e. $P(\mathscr{A})$ is nothing but the poset formed by all the sums of face ideals

---

[6]We consider this point of view since the same results are true if we consider the defining ideal of any arrangement of linear subspaces.

in the minimal primary decomposition of I ordered by reverse inclusion but notice that we have to identify these sums when they describe the same ideal.

Example 17. Consider the ideal $I = I_1 \cap I_2 \cap I_3 = (x_1, x_2) \cap (x_1, x_3) \cap (x_2, x_3)$ in $k[x_1, x_2, x_3]$. The initial pieces that allowed us to develop the Mayer–Vietoris process in the previous section can be encoded in the left poset. The poset $P(\mathscr{A})$ we have to consider now is the one on the right

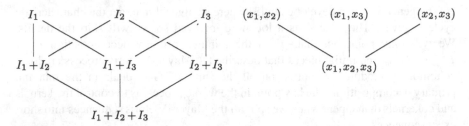

The Mayer–Vietoris spectral sequence has the following description when we consider the E_2-page:

$$E_2^{-i,j} = \varinjlim_{P(\mathscr{A})}{}^{(i)} H_{I_p}^j(R) \Rightarrow H_I^{j-i}(R)$$

where p is an element of the poset $P(\mathscr{A})$, I_p is the defining (radical) ideal of the irreducible variety corresponding to p and $\varinjlim_{P(\mathscr{A})}{}^{(i)}$ is the i-th left derived functor of the direct limit functor in the category of direct systems indexed by the poset $P(\mathscr{A})$.

On paper the E_2-page looks more difficult to deal with but we can give a nice topological interpretation. First recall that to any poset one can associate a simplicial complex, the so-called *order complex*, which has as vertices the elements of the poset and where a set of vertices $p_0, \ldots, p_r$ determines a r-dimensional simplex if $p_0 < \cdots < p_r$. Now we define $K(> p)$ to be the simplicial complex attached to the subposet $\{q \in P(\mathscr{A}) \mid q > p\}$ of $P(\mathscr{A})$. Then, there are R-module isomorphisms

$$\varinjlim_{P(\mathscr{A})}{}^{(i)} H_{I_p}^j(R) \simeq \bigoplus_{h(p)=j} [H_{I_p}^j(R) \otimes_k \tilde{H}_{i-1}(K(> p); k)],$$

where $\tilde{H}(-;k)$ denotes reduced simplicial homology and $h(p)$ denotes the height of the ideal I_p. Here we agree that the reduced homology with coefficients in k of the empty simplicial complex is k in degree -1 and zero otherwise.

The main result of this section is the following

Theorem 18. *Let $\mathscr{A}$ be an arrangement of linear varieties defined by a squarefree monomial ideal $I = I_1 \cap \cdots \cap I_m \subseteq R$. Then, the Mayer–Vietoris spectral sequence*

$$E_1^{-i,j} = \bigoplus_{1 \le \ell_1 < \cdots < \ell_i \le m} H_{I_{\ell_1}+\cdots+I_{\ell_i}}^j(R) \Longrightarrow H_I^{j-i}(R).$$

degenerates at the E_2-page.

The main ingredient of the proof is the fact that local cohomology modules supported on a face ideal only have one associated prime, in particular its characteristic variety only has one component. The degeneration of the Mayer–Vietoris spectral sequence provides a filtration of the local cohomology modules, where the successive quotients are given by the E_2-term.

Corollary 19. *Let $\mathscr{A}$ be an arrangement of linear varieties defined by a squarefree monomial ideal $I = I_1 \cap \cdots \cap I_n \subset R$. Then, for all $r \ge 0$ there is a filtration $\{F_j^r\}_{0 \le j \le n}$ of $H_I^r(R)$ by R-submodules such that*

$$F_j^r / F_{j-1}^r \cong \bigoplus_{h(p)=j} [\, H_{I_p}^j(R) \otimes_k \tilde{H}_{h(p)-r-1}(K(> p); k)\,].$$

Moreover, this is a filtration by holonomic D_R-modules when char$k = 0$. Thus we can compute the characteristic cycle of the modules $H_I^r(R)$ from the short exact sequences

$$(s_r): \qquad 0 \to F_{r-1}^r \to F_r^r \to F_r^r / F_{r-1}^r \to 0$$
$$(s_{r+1}): \qquad 0 \to F_r^r \to F_{r+1}^r \to F_{r+1}^r / F_r^r \to 0$$
$$\vdots$$
$$(s_n): \qquad 0 \to F_{n-1}^r \to H_I^r(R) \to F_n^r / F_{n-1}^r \to 0$$

given by the filtration and the additivity of the characteristic cycle. It is worth to point out that, by general properties of local cohomology modules, we have $F_j^r = 0$ $\forall j < r$.

Corollary 20. *The characteristic cycle of $H_I^r(R)$ is*

$$CC(H_I^r(R)) = \sum m_{n-r,p}\, T_{X_p}^* X,$$

where $m_{n-r,p} = \dim_k \tilde{H}_{h(p)-r-1}(K(> p); k)$.

Example 21. Consider $I = (x_1, x_2) \cap (x_1, x_3) \cap (x_2, x_3) \subseteq k[x_1, x_2, x_3]$. We have the poset

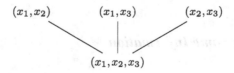

Then:

p	I_p	$K(> p)$	$\dim_k \tilde{H}_{-1}$	$\dim_k \tilde{H}_0$	$\dim_k \tilde{H}_1$
p_1	(x_1, x_2)	$\emptyset$	1	–	–
p_2	(x_1, x_3)	$\emptyset$	1	–	–
p_3	(x_2, x_3)	$\emptyset$	1	–	–
q	(x_1, x_2, x_3)	$\bullet\,\bullet\,\bullet$	–	2	–

And the characteristic cycle is

$$CC(H_I^2(R)) = T^*_{X_{(1,1,0)}} X + T^*_{X_{(1,0,1)}} X + T^*_{X_{(0,1,1)}} X + 2\ T^*_{X_{(1,1,1)}} X.$$

The poset associated to the next example has the same shape as the previous one but the corresponding characteristic cycle is quite different since the formula also depends on the height of the ideals in the poset.

Example 22. Consider the ideal $I = (x_1, x_2, x_5) \cap (x_3, x_4, x_5) \cap (x_1, x_2, x_3, x_4)$ in $k[x_1, x_2, x_3, x_4, x_5]$. The poset associated to this ideal is

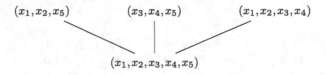

Therefore:

p	I_p	$K(> p)$	$\dim_k \tilde{H}_{-1}$	$\dim_k \tilde{H}_0$	$\dim_k \tilde{H}_1$
p_1	(x_1, x_2, x_5)	$\emptyset$	1	–	–
p_2	(x_3, x_4, x_5)	$\emptyset$	1	–	–
p_3	(x_1, x_2, x_3, x_4)	$\emptyset$	1	–	–
q	$(x_1, x_2, x_3, x_4, x_5)$	$\bullet\,\bullet\,\bullet$	–	2	–

In this case we have two local cohomology modules different from zero and their characteristic cycle are

$$CC(H_I^3(R)) = T^*_{X_{(1,1,0,0,1)}} X + T^*_{X_{(0,0,1,1,1)}} X,$$

$$CC(H_I^4(R)) = T^*_{X_{(1,1,1,1,0)}} X + 2\ T^*_{X_{(1,1,1,1,1)}} X.$$

2.7 Extracting Some Information

We have seen how to compute the characteristic cycle of the local cohomology module $H_I^r(R)$ using different techniques. This invariant describes the support of

$H_I^r(R)$ but we also get some extra information given by the multiplicities. In this section we will see how to extract information from the support but also, how the multiplicities can describe some arithmetic properties of the quotient ring R/I.

2.7.1 Support of Local Cohomology Modules

Once we know the characteristic cycle we can answer some of the questions raised by C. Huneke [109]. In particular we can deal with:

- Annihilation of local cohomology modules:
- Cohomological dimension.
- Description of the support of local cohomology modules.
- Krull dimension of local cohomology modules.
- Artinianity of local cohomology modules.

In the following example we can see how to read all this information from the coarser invariant given by the table $\Gamma(R/I)$ introduced in Sect. 1.8.

Example 23. Consider the ideal $I = (x_1, x_2, x_5) \cap (x_3, x_4, x_5) \cap (x_1, x_2, x_3, x_4)$ in $k[x_1, x_2, x_3, x_4, x_5]$. We have

$$CC(H_I^3(R)) = T^*_{X_{(1,1,0,0,1)}} X + T^*_{X_{(0,0,1,1,1)}} X,$$

$$CC(H_I^4(R)) = T^*_{X_{(1,1,1,1,0)}} X + 2\, T^*_{X_{(1,1,1,1,1)}} X.$$

Collecting the components by their dimension we get the table

$$\Gamma(R/I) = \begin{pmatrix} 0 & 2 & 0 \\ & 1 & 0 \\ & & 2 \end{pmatrix}.$$

- $\mathrm{Supp}(H_I^3(R)) = V(x_1, x_2, x_5) \cup V(x_3, x_4, x_5)$ $\dim_R \mathrm{Supp}_R H_I^3(R) = 2$.
- $\mathrm{Supp}(H_I^4(R)) = V(x_1, x_2, x_3, x_4)$ $\dim_R \mathrm{Supp}_R H_I^4(R) = 1$.
- $\mathrm{cd}\,(R, I) = 4$.

Notice that each column gives information on a different local cohomology module, in this case $H_I^5(R), H_I^4(R)$ and $H_I^3(R)$ respectively. Also, each row describe the dimension of the components of the characteristic cycle. Namely, the top row describe the components of dimension zero and so on.

2.7.2 Arithmetic Properties

The multiplicities of the characteristic cycle also provide a good test for the arithmetical properties of the quotient rings R/I.

- *Cohen–Macaulay property:* For a squarefree monomial ideal it is equivalent to have just one local cohomology module different from zero [128]. Therefore we only have to check out whether $\Gamma(R/I)$ has just one column.
- *Buchsbaum property:* By [182, Theorem 8.1], the Buchsbaum property of R/I is equivalent to the Cohen–Macaulayness of the localized rings $(R/I)_\mathfrak{p}$ for any prime ideal $\mathfrak{p} \neq \mathfrak{m}$. It means that the local cohomology modules $H_I^r(R)$ have dimension zero when $r \neq \operatorname{ht} I$, i.e. these modules are Artinian. Therefore, we have to check out that the non-vanishing entries of $\Gamma(R/I)$ are in the last column and the first row.
- *Gorenstein property:* This property is more involved and we have to check out the multiplicities. Namely, let $I = \mathfrak{p}_{\alpha_1} \cap \cdots \cap \mathfrak{p}_{\alpha_m}$ be the minimal primary decomposition of our squarefree monomial ideal. Then, R/I is Gorenstein if and only if R/I is Cohen–Macaulay and $m_{n-\operatorname{ht} I,\alpha} = 1$ for all $\alpha \geq \alpha_j$, $j = 1, \ldots, m$.

Example 24. Consider the ideals in $R = k[x_1, x_2, x_3, x_4]$:

- $I_1 = (x_1, x_2) \cap (x_3, x_4)$,
- $I_2 = (x_1, x_2) \cap (x_1, x_4) \cap (x_2, x_3) \cap (x_2, x_4)$,
- $I_3 = (x_1, x_2) \cap (x_1, x_4) \cap (x_2, x_3) \cap (x_3, x_4)$.

If we compute the corresponding characteristic cycles we get:

$$\Gamma(R/I_1) = \begin{pmatrix} 0 & 1 & 0 \\ & 0 & 0 \\ & & 2 \end{pmatrix}, \quad \Gamma(R/I_2) = \begin{pmatrix} 0 & 0 & 1 \\ & 0 & 4 \\ & & 4 \end{pmatrix}, \quad \Gamma(R/I_3) = \begin{pmatrix} 0 & 0 & 1 \\ & 0 & 4 \\ & & 4 \end{pmatrix}.$$

We have:

- R/I_1 is Buchsbaum but it is not Cohen–Macaulay.
- R/I_2 is Cohen–Macaulay but it is not Gorenstein.
- R/I_3 is Gorenstein.

If we take a look at the multiplicities we get:

$$CC(H_{I_2}^2(R)) = T^*_{X_{(1,1,0,0)}} X + T^*_{X_{(1,0,0,1)}} X + T^*_{X_{(0,1,1,0)}} X + T^*_{X_{(0,1,0,1)}} X +$$
$$+ T^*_{X_{(1,1,1,0)}} X + 2\, T^*_{X_{(1,1,0,1)}} X + T^*_{X_{(0,1,1,1)}} X + T^*_{X_{(1,1,1,1)}} X.$$

and

$$CC(H_{I_3}^2(R)) = T^*_{X_{(1,1,0,0)}} X + T^*_{X_{(1,0,0,1)}} X + T^*_{X_{(0,1,1,0)}} X + T^*_{X_{(0,0,1,1)}} X +$$
$$+ T^*_{X_{(1,1,0,1)}} X + T^*_{X_{(1,1,1,0)}} X + T^*_{X_{(1,0,1,1)}} X + T^*_{X_{(0,1,1,1)}} X +$$
$$+ T^*_{X_{(1,1,1,1)}} X.$$

2.7.3 Betti Numbers of Complements of Arrangements

Let $\mathscr{A}$ be an arrangement of linear varieties defined by a squarefree monomial ideal $I \subset R$. A formula for the Betti numbers of the complement $\mathbb{A}^n_{\mathbb{R}} - \mathscr{A}$ has been given by Goresky–MacPherson ([77, III.1.3. Theorem A]), which states (slightly reformulated) that

$$\tilde{H}_r(\mathbb{A}^n_{\mathbb{R}} - X; \mathbb{Z}) \cong \bigoplus_p \tilde{H}^{h(p)-r-2}(K(> p); \mathbb{Z}).$$

When we work over a field, these reduced simplicial cohomology groups allowed us to compute the characteristic cycle of the local cohomology modules so if we have

$$CC(H^r_I(R)) = \sum m_{n-r,p} T^*_{X_p} X$$

then, if $k = \mathbb{R}$ is the field of real numbers, the Betti numbers of the complement of the arrangement $\mathscr{A}$ in $X = \mathbb{A}^n_{\mathbb{R}}$ can be computed in terms of the multiplicities $\{m_{n-r,p}\}$ as

$$\dim_{\mathbb{Q}} \tilde{H}_r(\mathbb{A}^n_{\mathbb{R}} - \mathscr{A}; \mathbb{Q}) = \sum_p m_{n-(r+1),\, p}.$$

If $k = \mathbb{C}$ is the field of complex numbers, then one has

$$\dim_{\mathbb{Q}} \tilde{H}_r(\mathbb{A}^n_{\mathbb{C}} - \mathscr{A}; \mathbb{Q}) = \sum_p m_{n-(r+1-h(p)),\, p}.$$

Remark 25. Regarding a complex arrangement in $\mathbb{A}^n_{\mathbb{C}}$ as a real arrangement in $\mathbb{A}^{2n}_{\mathbb{R}}$, the formula for the Betti numbers of the complement of a complex arrangement follows from the formula for real arrangements.

Example 26. Consider the ideal $I = (x_1, x_2, x_5) \cap (x_3, x_4, x_5) \cap (x_1, x_2, x_3, x_4)$ in $k[x_1, x_2, x_3, x_4, x_5]$. We have

$$CC(H^3_I(R)) = T^*_{X_{(1,1,0,0,1)}} X + T^*_{X_{(0,0,1,1,1)}} X,$$

$$CC(H^4_I(R)) = T^*_{X_{(1,1,1,1,0)}} X + 2T^*_{X_{(1,1,1,1,1)}} X.$$

Therefore

$$\dim_{\mathbb{Q}} \tilde{H}_2(\mathbb{A}^5_{\mathbb{R}} - \mathscr{A}; \mathbb{Q}) = 2 \qquad \dim_{\mathbb{Q}} \tilde{H}_5(\mathbb{A}^5_{\mathbb{C}} - \mathscr{A}; \mathbb{Q}) = 2$$

$$\dim_{\mathbb{Q}} \tilde{H}_3(\mathbb{A}^5_{\mathbb{R}} - \mathscr{A}; \mathbb{Q}) = 3 \qquad \dim_{\mathbb{Q}} \tilde{H}_7(\mathbb{A}^5_{\mathbb{C}} - \mathscr{A}; \mathbb{Q}) = 1$$

$$\dim_{\mathbb{Q}} \tilde{H}_8(\mathbb{A}^5_{\mathbb{C}} - \mathscr{A}; \mathbb{Q}) = 2.$$

2.8   A General Framework: D-Modules with Variation Zero

Even though it provides a lot of information, the characteristic cycle does not describe completely the structure of the local cohomology modules. This fact is reflected in the work of A. Galligo et al. [72, 73] where they gave a description of the category of regular holonomic D_R-modules with support a normal crossing, e.g. local cohomology modules supported on squarefree monomial ideals, using the Riemann–Hilbert correspondence.

We will start considering the analytic situation where the Riemann–Hilbert correspondence takes place. For simplicity we will just consider the local situation where $R = \mathbb{C}\{x_1, \ldots, x_n\}$ is the ring of holomorphic functions in $X = \mathbb{C}^n$. The Riemann–Hilbert correspondence establishes an equivalence of categories between the category $\mathrm{Mod}_{hr}(D_R)$ of regular holonomic D_R-modules and the category $\mathrm{Perv}(\mathbb{C}^n)$ of perverse sheaves by means of the solutions functor $\mathbb{S}\mathrm{ol}(-) := \mathbb{R}\,\mathrm{Hom}_{D_R}(-, R)$.

Denote by $\mathrm{Perv}^T(\mathbb{C}^n)$ the subcategory of $\mathrm{Perv}(\mathbb{C}^n)$ of complexes of sheaves of finitely dimensional vector spaces on $\mathbb{C}^n$ which are perverse relatively to the given stratification of T [72, I.1], and by $\mathrm{Mod}_{hr}^T(D_R)$ the full abelian subcategory of the category of regular holonomic D_R-modules such that their solution complex is an object of $\mathrm{Perv}^T(\mathbb{C}^n)$. Then, the above equivalence gives by restriction an equivalence of categories between $\mathrm{Mod}_{hr}^T(D_R)$ and $\mathrm{Perv}^T(\mathbb{C}^n)$.

The category $\mathrm{Perv}^T(\mathbb{C}^n)$ has been described as a quiver representation in [72]. More precisely, they established an equivalence of categories with the category $\mathscr{C}^n$ whose objects are families $\{\mathscr{M}_\alpha\}_{\alpha \in \{0,1\}^n}$ of finitely dimensional $\mathbb{C}$-vector spaces, endowed with linear maps

$$\mathscr{M}_\alpha \xrightarrow{u_i} \mathscr{M}_{\alpha+\varepsilon_i} \ , \ \mathscr{M}_\alpha \xleftarrow{v_i} \mathscr{M}_{\alpha+\varepsilon_i}$$

for each $\alpha \in \{0,1\}^n$ such that $\alpha_i = 0$. These maps are called canonical (resp. variation) maps, and they are required to satisfy the conditions:

$$u_i u_j = u_j u_i, \quad v_i v_j = v_j v_i, \quad u_i v_j = v_j u_i \ \text{and} \ v_i u_i + id \ \text{is invertible.}$$

Such an object will be called an n-hypercube. A morphism between two n-hypercubes $\{\mathscr{M}_\alpha\}_\alpha$ and $\{\mathscr{N}_\alpha\}_\alpha$ is a set of linear maps $\{f_\alpha : \mathscr{M}_\alpha \to \mathscr{N}_\alpha\}_\alpha$, commuting with the canonical and variation maps.

Example 27. The 2-hypercube and the 3-hypercube. We follow the convention that the canonical maps u_i go downward and the variation maps v_i go upward:

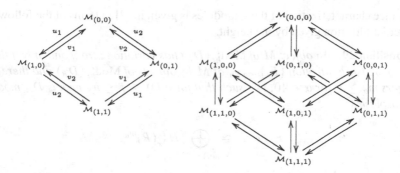

The construction of the n-hypercube corresponding to an object M of $\text{Mod}^T_{hr}(D_R)$ is explicitly given in [73]. We will skip the details because they are a little bit involved and depend on the solutions of our module M in some functional spaces. However we want to point out that the dimension of the $\mathbb{C}$-vector spaces $\mathcal{M}_\alpha$ in the n-hypercube are determined by the characteristic cycle of M. More precisely, if $CC(M) = \sum m_\alpha\, T^*_{X_\alpha} \mathbb{C}^n$ is the characteristic cycle of M, then for all $\alpha \in \{0,1\}^n$ one has the equality $\dim_{\mathbb{C}} \mathcal{M}_\alpha = m_\alpha$. It follows that the characteristic cycle is not enough to characterize a D_R-module with monomial support, i.e. support in T. We also need to describe the morphisms u_i's and v_i's. Therefore, the question that immediately pops up is:

Problem. How can we describe the n-hypercube associated to a local cohomology module $H^r_I(R)$ supported on a monomial ideal $I \subseteq R$?

To solve this question one needs to consider objects in the category $\text{Mod}^T_{hr}(D_R)$ having the following property (see [7, 8]):

Definition 28. We say that an object M of $\text{Mod}^T_{hr}(D_R)$ has variation zero if the morphisms v_i in the corresponding n-hypercube are zero for all $1 \le i \le n$ and all $\alpha \in \{0,1\}^n$ with $\alpha_i = 0$.

Modules with variation zero form a full abelian subcategory of $\text{Mod}^T_{hr}(D_R)$ but is not closed under extensions (see [6]). This category will be denoted $D^T_{v=0}$ and we will denote the corresponding category of n-hypercubes as $\mathscr{C}^n_{v=0}$. We have the following situation

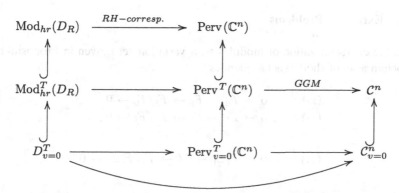

A nice characterization of these modules is given in [7] in terms of the following particular filtration given by the height.

Proposition 29. *An object M of $\mathrm{Mod}_{hr}^{T}(D_R)$ has variation zero if and only if there is an increasing filtration $\{F_j\}_{0 \leq j \leq n}$ of M by objects of $\mathrm{Mod}_{hr}^{T}(D_R)$ and there are integers $m_\alpha \geq 0$ for $\alpha \in \{0,1\}^n$, such that for all $0 \leq j \leq n$ one has D_R-module isomorphisms*

$$F_j/F_{j-1} \cong \bigoplus_{\substack{\alpha \in \{0,1\}^n \\ |\alpha|=j}} H_{\mathfrak{p}_\alpha}^{j}(R)^{m_\alpha}.$$

It follows from the degeneration of the Mayer–Vietoris spectral sequence (Corollary 19), that local cohomology modules supported on monomial ideals have variation zero.

Remark 30. The solutions of a regular holonomic D_R-modules are Nilsson class functions, i.e. are finite sums

$$f = \sum_{\beta,m} \varphi_{\beta,m}(\mathbf{x})(\log \mathbf{x})^m \mathbf{x}^\beta,$$

where $\varphi_{\beta,m}(\mathbf{x}) \in \mathbb{C}\{\mathbf{x}\}$, $\beta \in \mathbb{C}^n$ and $m \in (\mathbb{Z}^+)^n$ (see [21] for details). A Nilsson class function is a solution of a module with variation zero if and only if $m = \mathbf{0} \in \mathbb{Z}^n$ and $\beta \in \mathbb{Z}^n$, i.e. $f \in R_{x_1 \cdots x_n}$. This means, roughly speaking, that for a module with variation zero its solutions are algebraic.

It is easy to check out from its presentation

$$R_{\mathbf{x}^\alpha} \cong \frac{D_R}{D_R(\{x_i \partial_i + 1 \mid \alpha_i = 1\}, \{\partial_j \mid \alpha_j = 0\})}$$

that localizations at monomials are modules with variation zero. Therefore, using the Čech complex, we can also see that local cohomology modules supported on monomial ideals also belong to the category of modules with variation zero.

2.8.1 Extension Problems

From the characterization of modules with variation zero given in Proposition 29 we obtain a set of short exact sequences

$$
\begin{aligned}
(s_1): &\quad 0 \to F_0 \to F_1 \to F_1/F_0 \to 0 \\
(s_2): &\quad 0 \to F_1 \to F_2 \to F_2/F_1 \to 0 \\
&\qquad\qquad \vdots \\
(s_n): &\quad 0 \to F_{n-1} \to M \to F_n/F_{n-1} \to 0
\end{aligned}
$$

such that $F_j/F_{j-1} \cong \bigoplus_{|\alpha|=j} H^j_{\mathfrak{p}_\alpha}(R)^{m_\alpha}$. It follows from the additivity of the characteristic cycle that

$$CC(M) = \sum m_\alpha T^*_{X_\alpha} X.$$

The extension classes of this short exact sequences determine the structure of this module with variation zero so it is not enough considering the characteristic cycle. Recall that each extension class of the sequence (s_j) defines an element in $\mathrm{Ext}^1_{D^T_{v=0}}(F_j/F_{j-1}, F_{j-1})$.

Example 31. Assume that we have a module with variation zero M in a short exact sequence like

$$0 \longrightarrow H^2_{(x_1,x_2)}(R) \longrightarrow M \longrightarrow H^3_{(x_1,x_2,x_3)}(R) \longrightarrow 0.$$

Then, its characteristic cycle is $CC(M) = T^*_{X_{(1,1,0)}} X + T^*_{X_{(1,1,1)}} X$ but the short exact sequence may be split or not. In the first case we have the isomorphism $M \cong H^2_{(x_1,x_2)}(R) \oplus H^3_{(x_1,x_2,x_3)}(R)$ but in the second case we have that M is the injective module $E_{(1,1,0)}$ so is not isomorphic to the previous sum of local cohomology modules.

To determine a module with variation zero M we have to solve all the extensions problems associated to the corresponding filtration. On the other hand, to determine the corresponding n-hypercube we have to describe the linear maps u_i's. In Sect. 2.9 we will make this correspondence more precise with the help of the $\mathbb{Z}^n$-graded structure of M but, for the moment, the reader should notice the following correspondence:

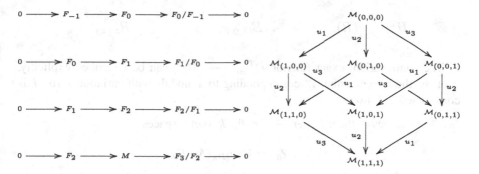

We will see that the extension problems at level j are uniquely determined by the canonical maps at the same level.

2.8.2 Modules with Variation Zero: The Algebraic Case

In principle we are only working in the analytic case but, since the solutions of a module with variation zero are algebraic, the equivalence $D_{v=0}^T \longrightarrow \mathscr{C}_{v=0}^n$ between modules with variation zero and the corresponding n-hypercubes can be extended to the algebraic case over any field of characteristic zero (see [8]). From now on, we will consider the polynomial ring $R = k[x_1, \ldots, x_n]$ over a field of characteristic zero, or even the formal power series ring $R = k[[x_1, \ldots, x_n]]$.

A very straightforward computation show us that in $D_{v=0}^T$ we have the following objects, $\forall \alpha \in \{0, 1\}^n$:

- *Simple:* $H_{\mathfrak{p}_\alpha}^{|\alpha|}(R) \cong \dfrac{R[\frac{1}{x^\alpha}]}{\sum_{\alpha_i=1} R[\frac{1}{x^{\alpha-\varepsilon_i}}]} \cong \dfrac{D_R}{D_R(\{x_i \mid \alpha_i=1\}, \{\partial_j \mid \alpha_j=0\})}.$
- *Injective:* $E_\alpha := \dfrac{R[\frac{1}{x}]}{\sum_{\alpha_i=1} R[\frac{1}{x^{1-\varepsilon_i}}]} \cong \dfrac{D_R}{D_R(\{x_i \mid \alpha_i=1\}, \{x_j \partial_j+1 \mid \alpha_j=0\})}.$
- *Projective:* $R_{x^\alpha} \cong \dfrac{D_R}{D_R(\{x_i \partial_i+1 \mid \alpha_i=1\}, \{\partial_j \mid \alpha_j=0\})}.$

Example 32. The 3-hypercube of a simple, injective and projective module with variation zero is as follows:

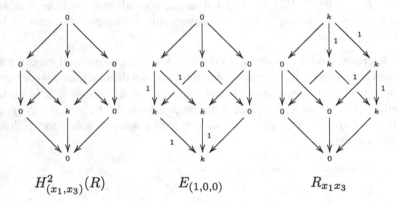

$$H_{(x_1, x_3)}^2(R) \qquad\qquad E_{(1,0,0)} \qquad\qquad R_{x_1 x_3}$$

The contravariant exact functor $D_{v=0}^T \longrightarrow \mathscr{C}_{v=0}^n$ can be described explicitly. Namely, the n-hypercube $\mathscr{M}$ corresponding to a module with variation zero M is constructed as follows:

(a) The vertices of the n-hypercube are the k-vector spaces

$$\mathscr{M}_\alpha := \mathrm{Hom}_{D_R}(M, E_\alpha).$$

(b) The linear maps u_i are induced by the epimorphisms $\pi_i : E_\alpha \to E_{\alpha+\varepsilon_i}$.

The positive characteristic case: Even though we do not have an analogue to the results of [72, 73] in positive characteristic, we can define a category of modules with variation zero and their n-hypercubes in positive characteristic. In this case one has to define modules with variation zero via the characterization given by the existence of an increasing filtration $\{F_j\}_{0 \le j \le n}$ of submodules of M such that

$$F_j / F_{j-1} \simeq \bigoplus_{|\alpha|=j} H^{|\alpha|}_{\mathfrak{p}_\alpha}(R)^{m_\alpha},$$

for some integers $m_\alpha \geq 0$, $\alpha \in \{0, 1\}^n$. Finally we point out that, using the same arguments as in [7, Lemma 4.4], the n-hypercube $\mathcal{M}$ associated to a module with variation zero M should be constructed using the following variant in terms of $\mathbb{Z}^n$-graded morphisms:

(a) The vertices of the n-hypercube are the k-vector spaces

$$\mathcal{M}_\alpha :=^* \mathrm{Hom}_R(M, E_\alpha).$$

(b) The linear maps u_i are induced by the epimorphisms $\pi_i : E_\alpha \to E_{\alpha+\varepsilon_i}$.

2.9 Building a Dictionary

To describe the D-module structure of local cohomology modules we introduced the category $D^T_{v=0}$ of modules with variation zero or equivalently, the category of n-hypercubes with variation zero $\mathscr{C}^n_{v=0}$. On the other hand, to study its $\mathbb{Z}^n$-graded structure we used the category **Str** of straight modules. In this section we will see that both approaches are equivalent.

On the other hand, we will also see that the study of local cohomology modules of a squarefree monomial ideal is equivalent to the study of free resolutions of the Alexander dual ideal. This equivalence gives a different point of view to the theory of free resolutions of monomial ideals since we can use some tools from the theory of local cohomology that are not in the realm of free resolutions such as the Čech complex, the Brodmann's sequence or any spectral sequence we use in this context. This equivalence will have a deepest insight in Sect. 3 where the Bass numbers of local cohomology modules can be interpreted in terms of the linear strands of the free resolution.

For the moment, we will make the following equivalences more precise:

Local cohomology $H^r_I(R)$		Free resolution $\mathbb{L}_\bullet(I^\vee)$
D-module structure	$\mathbb{Z}^n$-graded structure	$\mathbb{Z}^n$-graded structure
Characteristic cycle	$\mathbb{Z}^n$-graded pieces	$\mathbb{Z}^n$-graded Betti numbers
n-hypercubes	Morphisms between pieces	Linear strands

2.10 D-Module Structure vs. $\mathbb{Z}^n$-Graded Structure

At first sight we can find the analogies between the categories **Str**, $D^T_{v=0}$, $\mathscr{C}^n_{v=0}$ of straight modules, modules with variation zero and n-hypercubes respectively, since we characterized their objects as follows:

- *Straight modules*: A straight module is characterized by the $\mathbb{Z}^n$-graded pieces $M_{-\alpha}, \alpha \in \{0, 1\}^n$ and the multiplication by x_i maps $x_i : M_{-\alpha} \longrightarrow M_{-\alpha + \varepsilon_i}$.

  Equivalently, such a module comes with a filtration of $\mathbb{Z}^n$-graded submodules $\{F_j\}_{0 \le j \le n}$ such that

$$F_j / F_{j-1} \cong \bigoplus_{\substack{\alpha \in \{0,1\}^n \\ |\alpha| = j}} H_{\mathfrak{p}_\alpha}^j(\omega_R)^{m_\alpha}.$$

  Thus, in order to describe this module we have to solve the $\mathbb{Z}^n$-graded extension problems

$$(s_j): \qquad 0 \to F_{j-1} \to F_j \to F_j / F_{j-1} \to 0.$$

- *Modules with variation zero*: A module with variation zero is characterized by a filtration of D_R-modules $\{F_j\}_{0 \le j \le n}$ such that

$$F_j / F_{j-1} \cong \bigoplus_{\substack{\alpha \in \{0,1\}^n \\ |\alpha| = j}} H_{\mathfrak{p}_\alpha}^j(R)^{m_\alpha}.$$

  Then, to describe this module we have to solve the D_R-modules extension problems

$$(s_j): \qquad 0 \to F_{j-1} \to F_j \to F_j / F_{j-1} \to 0.$$

- *n-hypercubes*: A n-hypercube $\mathcal{M}$ is characterized by the pieces $\mathcal{M}_\alpha, \alpha \in \{0, 1\}^n$ and the canonical maps $u_i : \mathcal{M}_\alpha \longrightarrow \mathcal{M}_{\alpha + \varepsilon_i}$.

K. Yanagawa already proved in [200] that straight modules are D_R-modules so it will not come as a surprise that these modules, modulo a shift by **1**, have in fact variation zero. Recall that the local cohomology of the canonical module $H_I^r(\omega_R) = H_I^r(R(-\mathbf{1})) = H_I^r(R)(-\mathbf{1})$ supported on a monomial ideal I is straight but we only want to deal with $H_I^r(R)$ in the category of modules with variation zero.

Let $\varepsilon - \mathbf{Str}$ be the category of $\mathbb{Z}^n$-graded modules M such that $M(-\mathbf{1})$ is straight. An equivalence of categories $\varepsilon - \mathbf{Str} \longrightarrow D_{v=0}^T$ is established in [7, Theorem 4.3]. More precisely, a ε-straight module when viewed as a D_R-module has variation zero.

It is also proved in [7, Lemma 4.4] that given ε-straight modules M and N then we have functorial isomorphisms

$$^*\mathrm{Ext}_R^i(M, N) \cong \mathrm{Ext}_{D_{v=0}^T}^i(M, N)$$

for all $i \ge 0$. Therefore, the extension problems we had to solve to describe our module are equivalent in both categories. The following proposition is also proved in [7].

Proposition 33. *The extension class* (s_j) *is uniquely determined by the* k-*linear maps* $\cdot x_i : M_{-\alpha} \to M_{-\alpha+\varepsilon_i}$ *where* $|\alpha| = j$ *and* $\alpha_i = 1$.

Finally we will make the last remaining equivalence between straight modules and n-hypercubes more precise, i.e.

$$\varepsilon-\mathbf{Str} \xrightarrow{\hspace{3cm}} D^T_{v=0} \xrightarrow{\hspace{3cm}} \mathcal{C}^n_{v=0}$$

Let $M \in \varepsilon - \mathbf{Str}$ be a ε-straight module. The vertices and linear maps of corresponding n-hypercube $\mathcal{M} \in \mathcal{C}^n_{v=0}$ can be described from the $\mathbb{Z}^n$-graded pieces of M. Let $(M_{-\alpha})^*$ be the dual of the k-vector space defined by the piece of M of degree $-\alpha$, $\alpha \in \{0, 1\}^n$. Then, there are isomorphisms

$$\mathcal{M}_\alpha \cong (M_{-\alpha})^*$$

such that the following diagram commutes:

$$
\begin{array}{ccc}
\mathcal{M}_\alpha & \xrightarrow{\;u_i\;} & \mathcal{M}_{\alpha+\varepsilon_i} \\[2pt]
\cong \uparrow & & \cong \uparrow \\[2pt]
(M_{-\alpha})^* & \xrightarrow{\;(x_i)^*\;} & (M_{-\alpha-\varepsilon_i})^*
\end{array}
$$

where $(x_i)^*$ is the dual of the multiplication by x_i.

From now on we will loosely use the term *pieces of a module* M meaning the pieces of the n-hypercube associated to M but, if the reader is more comfortable with the $\mathbb{Z}^n$-graded point of view, one may also consider the $\mathbb{Z}^n$-graded pieces of M (with the appropriate sign).

For the case of local cohomology modules we can use M. Mustață's approach [146] to describe these linear maps. In this case we have the following commutative diagram:

$$
\begin{array}{ccc}
([H^r_I(R)]_{-\alpha})^* & \xrightarrow{\;(x_i)^*\;} & ([H^r_I(R)]_{-\alpha-\varepsilon_i})^* \\[4pt]
\cong \uparrow & & \uparrow \cong \\[4pt]
([H^r_I(R)])_\alpha & \xrightarrow{\;u_i\;} & ([H^r_I(R)])_{\alpha+\varepsilon_i} \\[4pt]
\cong \downarrow & & \downarrow \cong \\[4pt]
(\tilde{H}^{r-2}(\Delta^\vee_{1-\alpha}; k))^* & \xrightarrow{\;(v_i)^*\;} & (\tilde{H}^{r-2}(\Delta^\vee_{1-\alpha-\varepsilon_i}; k))^*
\end{array}
$$

where v_i is induced by the inclusion $\Delta^\vee_{1-\alpha} \subseteq \Delta^\vee_{1-\alpha-\varepsilon_i}$.

Example 34. Consider the ideal $I = (x_1x_2, x_1x_3, x_2x_3)$ in $R = k[x_1, x_2, x_3]$. The characteristic cycle of the local cohomology module describes the vertices of the corresponding 3-hypercube. Computing the linear maps among these vertices in an adequate basis we get:

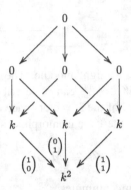

Remark 35. The advantage of the D-module approach is that it is more likely to be extended to other situations like the case of hyperplane arrangements. We recall that local cohomology modules with support an arrangement of linear subvarieties were already computed in [7] and a quiver representation of D_R-modules with support a hyperplane arrangement is given in [123, 124].

2.11 Local Cohomology vs. Free Resolutions

Let $R = k[x_1, \ldots, x_n]$ be the polynomial ring in n variables over a field k of any characteristic. Let $I = \mathfrak{p}_{\alpha_1} \cap \cdots \cap \mathfrak{p}_{\alpha_m}$ be the minimal primary decomposition of a squarefree monomial ideal. Its Alexander dual ideal $I^\vee$ is of the form $I^\vee = (\mathbf{x}^{\alpha_1}, \ldots, \mathbf{x}^{\alpha_m})$. The aim of this section is to relate the structure of the local cohomology modules $H_I^r(R)$ to the structure of the minimal free resolution of the ideal $I^\vee$.

2.11.1 Characteristic Cycle vs. Betti Numbers

M. Mustaţă [146, Corollary 3.1] already proved the following relation between the pieces of the local cohomology modules and the Betti numbers of the Alexander dual ideal

$$\beta_{j,\alpha}(I^\vee) = \dim_k [H_I^{|\alpha|-j}(R)]_\alpha$$

so the pieces of $H_I^r(R)$ for a fixed r describe the modules and the Betti numbers of the r-linear strand of $I^\vee$. Recall that $\dim_k [H_I^r(R)]_\alpha = m_{n-r,\alpha}$, where

$$CC(H_I^r(R)) = \sum m_{n-r,\alpha} T_{X_\alpha}^* X$$

is the characteristic cycle of the local cohomology module. Therefore we have:

Proposition 36. *Let* $I^\vee \subseteq R$ *be Alexander dual ideal of a squarefree monomial ideal* $I \subseteq R$. *Then we have:*

$$\beta_{j,\alpha}(I^\vee) = m_{n-|\alpha|+j,\alpha}(R/I).$$

Remark 37. As a summary, the multiplicities of the characteristic cycle $CC(H_I^r(R)) = \sum m_{n-r,\alpha} T_{X_\alpha}^* X$ can be described as follows:

$$
\begin{aligned}
m_{n-r,\alpha} &= \beta_{|\alpha|-r,\alpha}(I^\vee) \\
&= \dim_k [H_I^r(R)]_{-\alpha} \\
&= \dim_k \tilde{H}_{n-r-|\alpha|-1}(\mathrm{link}_\alpha \Delta; k) = \dim_k \tilde{H}^{r-2}(\Delta_{1-\alpha}^\vee; k) \\
&= \dim_k \tilde{H}_{r-2}(T_\alpha; k) \\
&= \dim_k \tilde{H}_{|\alpha|-r-1}(K(> \alpha); k) \\
&= \dim_k \mathrm{Hom}_{D_R}(M, E_\alpha) = \dim_k {}^* \mathrm{Hom}_R(M, E_\alpha)
\end{aligned}
$$

where the third equality comes from Terai's and Mustaţă's formula, the fourth comes from the computation given using the Čech complex, the fifth comes (slightly reformulated) from the Mayer–Vietoris spectral sequence and the last one is just the dimension of the corresponding piece in the n-hypercube.

The methods we used in Sect. 2.6 to compute the multiplicities of the characteristic cycle of local cohomology modules can be interpreted as follows:

- *Mayer–Vietoris process:* The initial pieces that we need to start the process are the local cohomology modules supported on sums of face ideals in the minimal primary decomposition $I = \mathfrak{p}_{\alpha_1} \cap \cdots \cap \mathfrak{p}_{\alpha_m}$. These sums are again face ideals and it is not difficult to check out that their Alexander duals are the least common multiples of the Alexander dual of each face ideal in the sum, i.e.

$$(\mathfrak{p}_{\alpha_{\ell_1}} + \cdots + \mathfrak{p}_{\alpha_{\ell_r}})^\vee = lcm(\mathbf{x}^{\alpha_{\ell_1}}, \cdots, \mathbf{x}^{\alpha_{\ell_r}}).$$

Thus, the information encoded by the initial pieces allow us to construct the Taylor resolution of the ideal $I^\vee$. The Mayer–Vietoris process can be understood as the process to pass from the Taylor resolution to a minimal free resolution of the ideal $I^\vee$.

- *Mayer–Vietoris spectral sequence:* The E_1-page of the Mayer–Vietoris spectral sequence also encodes the information needed to construct the Taylor resolution of the ideal $I^\vee$. If we take a close look to the E_2-page we will see that the poset $P(\mathscr{A})$ associated to the ideal I is nothing but the lcm-lattice of the Alexander dual ideal $I^\vee$. Therefore one recovers the formula for the Betti numbers given in [75].

Example 38. Let $I = (x_1, x_2) \cap (x_1, x_3) \cap (x_2, x_3) \subseteq k[x_1, x_2, x_3]$. The initial pieces that we use in the Mayer–Vietoris process that allow us to compute the characteristic cycle of the local cohomology modules $H_I^r(R)$ are

$$H_{(x_1,x_2)}^2(R), \qquad\qquad H_{(x_1,x_2,x_3)}^3(R), \qquad\qquad H_{(x_1,x_2,x_3)}^3(R)$$

$$H_{(x_1,x_3)}^2(R), \qquad\qquad H_{(x_1,x_2,x_3)}^3(R),$$

$$H_{(x_2,x_3)}^2(R), \qquad\qquad H_{(x_1,x_2,x_3)}^3(R),$$

corresponding to the sums of 1, 2 and 3 ideals in the minimal primary decomposition of I. Equivalently, these are the modules that appear in the E_1-page of the Mayer–Vietoris spectral sequence. The information given by these modules is equivalent to the information needed to describe the Taylor resolution of the Alexander dual ideal $I^\vee = (x_1 x_2, x_1 x_3, x_2 x_3)$.

$$
0 \longrightarrow R(-1,-1,-1) \longrightarrow
\begin{matrix} R(-1,-1,-1) \\ \oplus \\ R(-1,-1,-1) \\ \oplus \\ R(-1,-1,-1) \end{matrix}
\longrightarrow
\begin{matrix} R(-1,-1,0) \\ \oplus \\ R(-1,0,-1) \\ \oplus \\ R(0,-1,-1) \end{matrix}
\longrightarrow I^\vee \longrightarrow 0
$$

If we apply our Mayer–Vietoris process or compute the E_2-page of the Mayer–Vietoris spectral sequence we obtain the characteristic cycle

$$CC(H_I^2(R)) = T_{X_{(1,1,0)}}^* X + T_{X_{(1,0,1)}}^* X + T_{X_{(0,1,1)}}^* X + 2\, T_{X_{(1,1,1)}}^* X$$

that corresponds to the minimal free resolution

$$
0 \longrightarrow
\begin{matrix} R(-1,-1,-1) \\ \oplus \\ R(-1,-1,-1) \end{matrix}
\longrightarrow
\begin{matrix} R(-1,-1,0) \\ \oplus \\ R(-1,0,-1) \\ \oplus \\ R(0,-1,-1) \end{matrix}
\longrightarrow I^\vee \longrightarrow 0 \ .
$$

If R/I is Cohen–Macaulay then there is only one non vanishing local cohomology module so we can recover the following fundamental result of J.A. Eagon and V. Reiner [51].

Corollary 39. *Let $I^\vee \subseteq R$ be the Alexander dual ideal of a squarefree monomial ideal $I \subseteq R$. Then, R/I is Cohen–Macaulay if and only if $I^\vee$ has a linear free resolution.*

A generalization of this result expressed in terms of the projective dimension of R/I and the Castelnuovo–Mumford regularity of $I^\vee$ is given by N. Terai in [187]. We can also give a different approach by using the previous results.

Corollary 40. *Let $I^\vee \subseteq R$ be the Alexander dual ideal of a squarefree monomial ideal $I \subseteq R$. Then we have:*

$$\mathrm{pd}(R/I) = \mathrm{reg}(I^\vee).$$

Proof. By using Proposition 36 we have:

$$\mathrm{reg}(I^\vee) := \max\{|\alpha| - j \mid \beta_{j,\alpha}(I^\vee) \neq 0\} = \max\{|\alpha| - j \mid m_{n-|\alpha|+j,\alpha}(R/I) \neq 0\}.$$

Then, by [3, Corollary 3.13] we get the desired result since:

$$\mathrm{pd}(R/I) = \mathrm{cd}(R, I) = \max \{|\alpha| - j \mid m_{n-|\alpha|+j,\alpha}(R/I) \neq 0\},$$

where the first assertion comes from [128]. $\square$

Mayer–Vietoris splittings: Splittings of a monomial ideal have a long tradition in the study of free resolutions (see [56, 60, 83]). One looks for a decomposition $I = J + K$ of our ideal satisfying the following formula for the $\mathbb{Z}^n$-graded Betti numbers

$$\beta_{i,\alpha}(I) = \beta_{i,\alpha}(J) + \beta_{i,\alpha}(K) + \beta_{i-1,\alpha}(J \cap K).$$

C. Francisco, H. T. Hà and A. Van Tuyl coined the term Betti splitting in [65] to tackle this formula. Using our approach it is easy to check out that the condition of being a Betti splitting is nothing but the splitting of the Mayer–Vietoris exact sequence for the local cohomology modules of its Alexander dual $I^\vee = J^\vee \cap K^\vee$

$$\cdots \longrightarrow H^r_{J^\vee+K^\vee}(R) \longrightarrow H^r_{J^\vee}(R) \oplus H^r_{K^\vee}(R) \longrightarrow H^r_{I^\vee}(R) \longrightarrow H^{r+1}_{J^\vee+K^\vee}(R) \longrightarrow \cdots$$

Definition 41. We say that a squarefree monomial ideal $I^\vee = J^\vee \cap K^\vee$ is r-MV-splittable if the corresponding Mayer–Vietoris exact sequence splits at level r, i.e. we have a short exact sequence

$$0 \longrightarrow H^r_{J^\vee}(R) \oplus H^r_{K^\vee}(R) \longrightarrow H^r_{I^\vee}(R) \longrightarrow H^{r+1}_{J^\vee+K^\vee}(R) \longrightarrow 0.$$

A squarefree monomial ideal $I^\vee$ is MV-splittable if it is r-MV-splittable for all r.

The following result comes easily from Proposition 36

Proposition 42. *A squarefree monomial ideal I is Betti-splittable with splitting $I = J + K$ if and only if $I^\vee = J^\vee \cap K^\vee$ is MV-splittable. In this case, for all $i \geq 0$, $\alpha \in \{0, 1\}^n$*

$$\beta_{i,\alpha}(I) = \beta_{i,\alpha}(J) + \beta_{i,\alpha}(K) + \beta_{i-1,\alpha}(J \cap K).$$

The condition of $I^\vee$ being r-MV-splittable is equivalent to have the same formula for the Betti numbers in the r-linear strand.

2.11.2 *n*-Hypercubes vs. Linear Strands

First, notice that, giving the appropriate sign to the canonical maps of the hypercube $\mathscr{M} = \{\mathscr{M}_\alpha\}_{\alpha \in \{0,1\}^n}$ associated to a module with variation zero M, we can construct the following complex of k-vector spaces:

$$\mathscr{M}^\bullet : 0 \longleftarrow \mathscr{M}_1 \overset{u_0}{\longleftarrow} \bigoplus_{|\alpha|=n-1} \mathscr{M}_\alpha \overset{u_1}{\longleftarrow} \cdots \overset{u_{p-1}}{\longleftarrow} \bigoplus_{|\alpha|=n-p} \mathscr{M}_\alpha \overset{u_p}{\longleftarrow} \cdots \overset{u_{n-1}}{\longleftarrow} \mathscr{M}_0 \longleftarrow 0$$

where the map between summands $\mathscr{M}_\alpha \longrightarrow \mathscr{M}_{\alpha+\varepsilon_i}$ is $\mathrm{sign}(i, \alpha + \varepsilon_i)$ times the canonical map $u_i : \mathscr{M}_\alpha \longrightarrow \mathscr{M}_{\alpha+\varepsilon_i}$.

Example 43. Three-hypercube and its associated complex

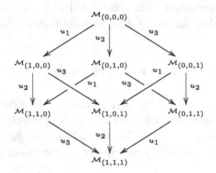

Our main result in this section is that the matrices in the complex of k-vector spaces associated to the n-hypercube of a fixed local cohomology module $H_I^r(R)$ are the transpose of the monomial matrices of the r-linear strand[7] of the Alexander dual ideal $I^\vee$. To prove the following proposition one has to put together some results scattered in the work of K. Yanagawa [199, 200].

---

[7]In the language of [157] we would say that the n-hypercube has the same information as the frame of the r-linear strand.

Proposition 44. *Let* $\mathcal{M} = \{[H_I^r(R)]_\alpha\}_{\alpha\in\{0,1\}^n}$ *be the n-hypercube of a fixed local cohomology module* $H_I^r(R)$ *supported on a monomial ideal* $I \subseteq R = k[x_1,\ldots,x_n]$. *Then,* $\mathcal{M}^\bullet$ *is the complex of k-vector spaces whose matrices are the transpose of the monomial matrices of the r-linear strand* $\mathbb{L}_\bullet^{<r>}(I^\vee)$ *of the Alexander dual ideal of* I.

Proof. Given an squarefree module M, K. Yanagawa constructed in [199] a chain complex $\mathbb{F}_\bullet(M)$ of free R-modules as follows:

$$\mathbb{F}_\bullet(M):$$

$$0 \longrightarrow [M]_1 \otimes_k R \xrightarrow{d_0} \cdots \xrightarrow{d_{p-1}} \bigoplus_{|\alpha|=n-p} [M]_\alpha \otimes_k R \xrightarrow{d_p} \cdots \xrightarrow{d_{n-1}} [M]_0 \otimes_k R \longrightarrow 0$$

where the map between summands $[M]_{\alpha+\varepsilon_i} \otimes_k R \longrightarrow [M]_\alpha \otimes_k R$ sends $y \otimes 1 \in [M]_{\alpha+\varepsilon_i} \otimes_k R$ to $\mathrm{sign}(i, \alpha + \varepsilon_i)\,(x_i\, y \otimes x_i)$. For the particular case of $M = \mathrm{Ext}_R^r(R/I, R(-1))$ he proved an isomorphism (after an appropriate shifting) between $\mathbb{F}_\bullet(M)$ and the r-linear strand $\mathbb{L}_\bullet^{<r>}(I^\vee)$ of the Alexander dual ideal $I^\vee$ of I (see [199, Theorem 4.1]).

We have an equivalence of categories between squarefree modules and straight modules [200, Proposition 2.8], thus one may also construct the chain complex $\mathbb{F}_\bullet(M)$ for any straight module M. By this equivalence, the squarefree module $\mathrm{Ext}_R^r(R/I, R(-1))$ corresponds to the local cohomology module $H_I^r(R)(-1)$ so there is an isomorphism between $\mathbb{F}_\bullet(H_I^r(R)(-1))$ and the r-linear strand $\mathbb{L}_\bullet^{<r>}(I^\vee)$ after an appropriate shifting. Taking a close look to the construction of $\mathbb{F}_\bullet(M)$ one may check that the scalar entries in the corresponding monomial matrices are obtained by transposing the scalar entries in the one associated to the hypercube of $H_I^r(R)$ with the appropriate shift. More precisely, if

$$\mathbb{L}_\bullet^{<r>}(I^\vee): \quad 0 \longrightarrow L_{n-r}^{<r>} \longrightarrow \cdots \longrightarrow L_1^{<r>} \longrightarrow L_0^{<r>} \longrightarrow 0\,,$$

is the r-linear strand of the Alexander dual ideal $I^\vee$ then we transpose its monomial matrices to obtain a complex of k-vector spaces indexed as follows:

$$\mathbb{F}_\bullet^{<r>}(I^\vee)^*: \quad 0 \longleftarrow K_0^{<r>} \longleftarrow \cdots \longleftarrow K_{n-r-1}^{<r>} \longleftarrow K_{n-r}^{<r>} \longleftarrow 0\,.$$

3 Bass Numbers of Local Cohomology Modules

Throughout this section we will consider either the polynomial ring $R = [x_1,\ldots,x_n]$ or the formal power series ring $R = [[x_1,\ldots,x_n]]$ over a field k. Given a squarefree monomial ideal $I \subseteq R$ we are going to give a method to compute Bass numbers of local cohomology modules $H_I^r(R)$, in particular Lyubeznik numbers.

These local cohomology modules are modules with variation zero so we are going to work in this framework and study Bass numbers of such modules.

An interesting outcome of the dictionary between local cohomology modules and free resolutions we developed in the previous section is that Lyubeznik numbers can be interpreted as a measure of the acyclicity of the linear strands of the Alexander dual ideal. This fact opens up a number of questions around this invariant.

Later on we will turn our attention to the injective resolution of local cohomology modules. To describe completely these resolutions is out of the scope of these work, but we can relate the behavior of Bass numbers by using the structure, i.e. the n-hypercube, of the local cohomology modules. In particular we can give a bound for the injective dimension.

3.1 Bass Numbers of Modules with Variation Zero

Let $M \in D_{v=0}^{T}$ be a module with variation zero. The aim of this section is to compute the pieces of the local cohomology module $H_{\mathfrak{p}_\alpha}^p(M)$, for any given homogeneous prime ideal $\mathfrak{p}_\alpha$, $\alpha \in \{0, 1\}^n$. This module also belongs to $D_{v=0}^T$ so we want to compute the pieces of the corresponding n-hypercube $\{[H_{\mathfrak{p}_\alpha}^p(M)]_\beta\}_{\beta \in \{0,1\}^n} \in \mathscr{C}_{v=0}^n$. Among these pieces we find the Bass numbers of M (see [6, Proposition 4.1]). Namely, we have

$$\mu_p(\mathfrak{p}_\alpha, M) = \dim_k [H_{\mathfrak{p}_\alpha}^p(M)]_\alpha.$$

Bass numbers have a good behavior with respect to localization so we can always assume that $\mathfrak{p}_\alpha = \mathfrak{m}$ is the maximal ideal and $\mu_p(\mathfrak{m}, M) = \dim_k [H_{\mathfrak{m}}^p(M)]_1$.

Remark 45. Let $\mathscr{M} \in \mathscr{C}_{v=0}^n$ be an n-hypercube. The restriction of $\mathscr{M}$ to a face ideal $\mathfrak{p}_\alpha$, $\alpha \in \{0, 1\}^n$ is the $|\alpha|$-hypercube $\mathscr{M}_{\leq\alpha} := \{\mathscr{M}_\beta\}_{\beta \leq \alpha} \in \mathscr{C}_{v=0}^{|\alpha|}$ (see [6, Proposition 3.1]). This gives a functor that in some cases plays the role of the localization functor. In particular, to compute the Bass numbers with respect to $\mathfrak{p}_\alpha$ of a module with variation zero M we only have to consider the corresponding $|\alpha|$-hypercube $\mathscr{M}_{\leq\alpha}$ so we may assume that $\mathfrak{p}_\alpha$ is the maximal ideal.

Later on we will specialize to the case of M being a local cohomology module $H_I^r(R)$. In particular, we will give a different approach to the computation of the Bass numbers of these modules given by K. Yanagawa in [200] or the algorithmic computation given by the author in [3, 4].

3.2 Computing Bass Numbers

The degree **1** part of the hypercube corresponding to the local cohomology module $H_{\mathfrak{m}}^p(M)$ is the p-th homology of the complex of k-vector spaces

$[\check{C}_{\mathfrak{m}}(M)]_1^\bullet$:

$$0 \longleftarrow [M]_1 \xleftarrow{\overline{d_0}} \bigoplus_{|\alpha|=1} [M_{x^\alpha}]_1 \xleftarrow{\overline{d_1}} \cdots \xleftarrow{\overline{d_{p-1}}} \bigoplus_{|\alpha|=p} [M_{x^\alpha}]_1 \xleftarrow{\overline{d_p}} \cdots \xleftarrow{\overline{d_{n-1}}} [M_{x^1}]_1 \longleftarrow 0$$

that we obtain applying the exact functor[8] $\mathrm{Hom}_{D_R}(\cdot, E_1)$ to the Čech complex

$$\check{C}_{\mathfrak{m}}^\bullet(M) : 0 \longrightarrow M \xrightarrow{d_0} \bigoplus_{|\alpha|=1} M_{x^\alpha} \xrightarrow{d_1} \cdots \xrightarrow{d_{p-1}} \bigoplus_{|\alpha|=p} M_{x^\alpha} \xrightarrow{d_p} \cdots \xrightarrow{d_{n-1}} M_{x^1} \longrightarrow 0,$$

On the other hand, associated to the hypercube $\mathcal{M} = \{\mathcal{M}_\alpha\}_{\alpha \in \{0,1\}^n}$ we constructed a complex of k-vector spaces:

$$\mathcal{M}^\bullet : 0 \longleftarrow \mathcal{M}_1 \xleftarrow{u_0} \bigoplus_{|\alpha|=n-1} \mathcal{M}_\alpha \xleftarrow{u_1} \cdots \xleftarrow{u_{p-1}} \bigoplus_{|\alpha|=n-p} \mathcal{M}_\alpha \xleftarrow{u_p} \cdots \xleftarrow{u_{n-1}} \mathcal{M}_0 \longleftarrow 0.$$

Proposition 46. *Let $M \in D_{v=0}^T$ be a module with variation zero and $\mathcal{M}^\bullet$ its corresponding complex associated to the n-hypercube. Then, there is an isomorphism of complexes $\mathcal{M}^\bullet \cong [\check{C}_{\mathfrak{m}}(M)]_1^\bullet$. In particular $[H_{\mathfrak{m}}^p(M)]_1 \cong H_p(\mathcal{M}^\bullet)$.*

Therefore we have the following characterization of Bass numbers:

Corollary 47. *Let $M \in D_{v=0}^T$ be a module with variation zero and $\mathcal{M}^\bullet$ its corresponding complex associated to the n-hypercube. Then*

$$\mu_p(\mathfrak{m}, M) = \dim_k H_p(\mathcal{M}^\bullet).$$

Example 48. Let $R = k[x_1, x_2, x_3]$. Consider the 3-hypercube

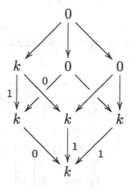

of a random module $M \in D_{v=0}^T$. It is not difficult to check out that in fact

---

[8]In positive characteristic we apply the functor $^*\mathrm{Hom}_R(\cdot, E_1)$.

$$M \cong (H^1_{(x_1)}(R))_{x_2} \oplus H^2_{(x_1 x_2, x_3)}(R)$$

but this is not important for our purposes. We are going to compute the Bass numbers with respect to any homogeneous prime ideal $\mathfrak{p}_\alpha$, $\alpha \in \{0,1\}^3$.

- For $\alpha = (1,1,1)$, the complex $\mathscr{M}^\bullet$ associated to the 3-hypercube is:

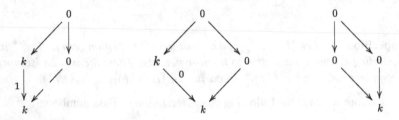

$$0 \longleftarrow k \xleftarrow{\;(0,1,1)\;} k^3 \xleftarrow{\;\begin{pmatrix} -1 \\ 0 \\ 0 \end{pmatrix}\;} k \longleftarrow 0 \longleftarrow 0$$

Therefore $\mu_1(\mathfrak{m}, M) = \dim_k H_1(\mathscr{M}^\bullet) = 1$ and $\mu_p(\mathfrak{m}, M) = 0$ for all $p \neq 1$.
- The restriction of $\mathscr{M}$ to the face ideals $\mathfrak{p}_\alpha$ for $\alpha = (1,1,0), (1,0,1)$ and $(0,1,1)$ are respectively

- The restriction of $\mathscr{M}$ to the face ideals $\mathfrak{p}_\alpha$ for $\alpha = (1,0,0), (0,1,0)$ and $(0,0,1)$ are

After computing the homology of the associated complexes of k-vector spaces, the Bass numbers of M are:

$\mathfrak{p}_\alpha$	μ_0	μ_1	μ_2
(x_1)	1	–	–
(x_2)	–	–	–
(x_3)	–	–	–
(x_1, x_2)	–	–	–
(x_1, x_3)	1	1	–
(x_2, x_3)	1	–	–
(x_1, x_2, x_3)	–	1	–

3.2.1 Extracting Some Information

Once we compute Bass numbers of modules with variation zero we can deal with the following questions:

- Annihilation of Bass numbers.
- Injective dimension.
- Description of the small support of local cohomology modules.
- Associated primes of local cohomology modules.

Example 49. Consider the module with variation zero in the previous example. Notice that we have $\mathrm{Supp}_R M = V(x_1) \cup V(x_2, x_3)$, so $\dim_R \mathrm{Supp}_R M = 2$. In this case, the inclusion $\mathrm{supp}_R M \subset \mathrm{Supp}_R M$ is strict since (x_1, x_2) does not belong to the small support.

On the other hand, the $\mathbb{Z}^3$-graded injective dimension is $^*\mathrm{id}_R M = 1$. Finally, the set of associated primes, i.e. those prime ideals having 0-th Bass number different from zero are

$$\mathrm{Ass}_R M = \{(x_1), (x_1, x_3), (x_2, x_3)\}.$$

In particular (x_1, x_3) is an associated prime that is not minimal in the support of M.

3.3 Computing Lyubeznik Numbers

Now we specialize to the case where our module with variation zero is a local cohomology module $H_I^r(R)$ supported on a monomial ideal $I \subseteq R$. We want to compute

$$\lambda_{p,n-r}(R/I) = \mu_p(\mathfrak{m}, H_I^r(R)) = \dim_k [H_{\mathfrak{m}}^p(H_I^r(R))]_1.$$

Remark 50. Our approach is different form [200] where K. Yanagawa gave the following formula for Lyubeznik numbers:

$$\lambda_{p,n-r}(R/I) = \dim_k [\mathrm{Ext}_R^{n-p}(\mathrm{Ext}_R^{n-r}(R/I, R), R)]_0.$$

Let $\mathcal{M} = \{[H_I^r(R)]_\alpha\}_{\alpha \in \{0,1\}^n}$ be the n-hypercube corresponding to $H_I^r(R)$. In this case we have a topological description of the pieces and linear maps of the n-hypercube using M. Mustaţă's approach [146]. Thus, the complex of k-vector spaces associated to $\mathcal{M}$ is:

$$\mathcal{M}^\bullet : 0 \longleftarrow \tilde{H}^{r-2}(\Delta_0^\vee; k) \overset{u_0}{\longleftarrow} \cdots \overset{u_{p-1}}{\longleftarrow} \bigoplus_{|\alpha|=p} \tilde{H}^{r-2}(\Delta_\alpha^\vee; k) \overset{u_p}{\longleftarrow} \cdots \overset{u_{n-1}}{\longleftarrow} \tilde{H}^{r-2}(\Delta_1^\vee; k) \longleftarrow 0$$

where the map between summands $\tilde{H}^{r-2}(\Delta_{\alpha+\varepsilon_i}^\vee; k) \longrightarrow \tilde{H}^{r-2}(\Delta_\alpha^\vee; k)$, is induced by the inclusion $\Delta_\alpha^\vee \subseteq \Delta_{\alpha+\varepsilon_i}^\vee$. In particular, the Lyubeznik numbers of R/I are

$$\lambda_{p,n-r}(R/I) = \dim_k H_p(\mathscr{M}^\bullet).$$

At this point one may wonder whether there is a simplicial complex, a regular cell complex, or a CW-complex that supports the complex of k-vector spaces $\mathscr{M}^\bullet$ so one may get a Hochster-like formula not only for the pieces of the local cohomology modules $H_I^r(R)$ but for its Bass numbers as well. Unfortunately this is not the case in general. To check this out we will use our dictionary developed in Sect. 2.9 to make a detour through the theory of free resolutions of monomial ideals. Then we refer to the work of M. Velasco [191] to find examples of free resolutions that are not supported by CW-complexes.

3.3.1 An Interpretation of Lyubeznik Numbers

Let $\mathscr{M} = \{[H_I^r(R)]_\alpha\}_{\alpha \in \{0,1\}^n}$ be the n-hypercube associated to a fixed local cohomology module $H_I^r(R)$ supported on a monomial ideal $I \subseteq R = k[x_1, \ldots, x_n]$
    Consider the r-linear strand of the Alexander dual ideal $I^\vee$

$$\mathbb{L}_\bullet^{<r>}(I^\vee): \quad 0 \longrightarrow L_{n-r}^{<r>} \longrightarrow \cdots \longrightarrow L_1^{<r>} \longrightarrow L_0^{<r>} \longrightarrow 0 .$$

Transposing its monomial matrices we obtain a complex of k-vector spaces

$$\mathbb{F}_\bullet^{<r>}(I^\vee)^*: \quad 0 \longleftarrow K_0^{<r>} \longleftarrow \cdots \longleftarrow K_{n-r-1}^{<r>} \longleftarrow K_{n-r}^{<r>} \longleftarrow 0$$

that is isomorphic to the complex of k-vector spaces $\mathscr{M}^\bullet$ associated to the n-hypercube. Then we have:

Corollary 51. *Let $\mathbb{F}_\bullet^{<r>}(I^\vee)^*$ be the complex of k-vector spaces obtained from the r-linear strand of the minimal free resolution of the Alexander dual ideal $I^\vee$ transposing its monomial matrices. Then*

$$\lambda_{p,n-r}(R/I) = \dim_k H_p(\mathbb{F}_\bullet^{<r>}(I^\vee)^*).$$

It follows that one may think Lyubeznik numbers of a squarefree monomial I as a measure of the acyclicity of the r-linear strand of the Alexander dual $I^\vee$.

Remark 52. As a summary of the dictionary between local cohomology modules and free resolutions we have:

- The pieces $[H_I^r(R)]_\alpha$ correspond to the Betti numbers $\beta_{|\alpha|-r,\alpha}(I^\vee)$.
- The n-hypercube of $H_I^r(R)$ corresponds to the r-linear strand $\mathbb{L}_\bullet^{<r>}(I^\vee)$.

Given a free resolution $\mathbb{L}_\bullet$ of a finitely generated graded R-module M, D. Eisenbud et al. [53] defined its *linear part* as the complex $\lin(\mathbb{L}_\bullet)$ obtained by erasing the terms of degree ≥ 2 from the matrices of the differential maps. To measure the acyclicity of the linear part, J. Herzog and S. Iyengar [94] introduced the *linearity defect* of M as $\mathrm{ld}_R(M) := \sup\{p \mid H_p(\lin(\mathbb{L}_\bullet)) \neq 0\}$. Therefore we also have:

- The n-hypercubes of $H_I^r(R)$, $\forall r$ correspond to the linear part $\lin(\mathbb{L}_\bullet(I^\vee))$.
- The Lyubeznik table of R/I can be viewed as a generalization of $\mathrm{ld}_R(I^\vee)$.

3.3.2 Examples

It is well-known that Cohen–Macaulay squarefree monomial ideals have a trivial Lyubeznik table

$$\Lambda(R/I) = \begin{pmatrix} 0 \cdots 0 \\ \ddots \vdots \\ 1 \end{pmatrix}$$

because they only have one non-vanishing local cohomology module. Recall that its Alexander dual has a linear resolution (see [51, Theorem 3]) so its acyclic. In general, there are non-Cohen–Macaulay ideals with trivial Lyubeznik table. Some of them are far from having only one local cohomology module different from zero.

Example 53. Consider the ideal in $k[x_1, \ldots, x_9]$:

$$I = (x_1, x_2) \cap (x_3, x_4) \cap (x_5, x_6) \cap (x_7, x_8) \cap (x_9, x_1) \cap (x_9, x_2) \cap (x_9, x_3) \cap$$

$$\cap (x_9, x_4) \cap (x_9, x_5) \cap (x_9, x_6) \cap (x_9, x_7) \cap (x_9, x_8).$$

The non-vanishing local cohomology modules are $H_I^r(R)$, $r = 2, 3, 4, 5$ but the Lyubeznik table is trivial.

Ideals with trivial Lyubeznik table can be characterized by the acyclicity of the linear strands.

Proposition 54. *Let* $I \subseteq R = k[x_1, \ldots, x_n]$ *be a squarefree monomial ideal. Then, the following conditions are equivalent:*

i) R/I *has a trivial Lyubeznik table.*
ii) $H_{(n-r)-i}(\mathbb{F}_\bullet^{<r>}(I^\vee)^*) = 0 \;\; \forall i > 0$ *if* $r = \mathrm{ht}\, I$ *and*
$ H_{(n-r)-i}(\mathbb{F}_\bullet^{<r>}(I^\vee)^*) = 0 \;\; \forall i \geq 0$ *if* $r \neq \mathrm{ht}\, I$.

Notice that the second condition is close to $I^\vee$ being *componentwise linear*. This notion was introduced by J. Herzog and T. Hibi in [88] where they also proved that their Alexander dual belong to the class of *sequentially Cohen–Macaulay* ideals given by R. Stanley [182]. On the other hand, K. Yanagawa [199, Proposition 4.9] and T. Römer [167, Theorem 3.2.8] characterized componentwise linear ideals as those having acyclic linear strands in homological degree different from zero. Namely, the ideal $I^\vee$ is componentwise linear if and only if $H_i(\mathbb{L}_\bullet^{<r>}(I^\vee)) = 0$ $\forall i > 0$ and $\forall r$. The previous example has a trivial Lyubeznik table but R/I is not sequentially Cohen–Macaulay.

The simplest examples of ideals with non-trivial Lyubeznik table are minimal non-Cohen–Macaulay squarefree monomial ideals (see [131])

Example 55. The unique minimal non-Cohen–Macaulay squarefree monomial ideal of pure height two in $R = k[x_1, \ldots, x_n]$ is:

$$\mathfrak{a}_n = (x_1, x_3) \cap \cdots \cap (x_1, x_{n-1}) \cap (x_2, x_4) \cap \cdots \cap (x_2, x_n) \cap \cdots \cap (x_{n-2}, x_n).$$

- $\mathfrak{a}_4 = (x_1, x_3) \cap (x_2, x_4)$.

We have $H^2_{\mathfrak{a}_4}(R) \cong H^2_{(x_1, x_3)}(R) \oplus H^2_{(x_2, x_4)}(R)$ and $H^3_{\mathfrak{a}_4}(R) \cong E_1$. Thus its Lyubeznik table is

$$\Lambda(R/\mathfrak{a}_4) = \begin{pmatrix} 0 & 1 & 0 \\ & 0 & 0 \\ & & 2 \end{pmatrix}.$$

- $\mathfrak{a}_5 = (x_1, x_3) \cap (x_1, x_4) \cap (x_2, x_4) \cap (x_2, x_5) \cap (x_3, x_5)$.

We have $H^3_{\mathfrak{a}_5}(R) \cong E_1$ and the hypercube associated to $H^2_{\mathfrak{a}_5}(R)$ satisfy $[H^2_{\mathfrak{a}_5}(R)]_\alpha \cong k$ for

- $\alpha = (1,0,1,0,0), (1,0,0,1,0), (0,1,0,1,0), (0,1,0,0,1), (0,0,1,0,1)$
- $\alpha = (1,1,0,1,0), (1,0,1,1,0), (1,0,1,0,1), (0,1,1,0,1), (0,1,0,1,1)$

The complex associated to the hypercube is

$$0 \longleftarrow 0 \longleftarrow 0 \longleftarrow k^5 \overset{u_2}{\longleftarrow} k^5 \longleftarrow 0 \longleftarrow 0 \longleftarrow 0$$

where the matrix corresponding to u_2 is the rank 4 matrix:

$$\begin{pmatrix} 0 & -1 & -1 & 0 & 0 \\ 1 & -1 & 0 & 0 & 0 \\ -1 & 0 & 0 & 0 & 1 \\ 0 & 0 & 0 & 1 & 1 \\ 0 & 0 & -1 & -1 & 0 \end{pmatrix}.$$

Thus its Lyubeznik table is

$$\Lambda(R/\mathfrak{a}_5) = \begin{pmatrix} 0 & 0 & 1 & 0 \\ & 0 & 0 & 0 \\ & & 0 & 1 \\ & & & 1 \end{pmatrix}.$$

In general one gets

$$\Lambda(R/\mathfrak{a}_n) = \begin{pmatrix} 0\,0\,0\,\cdots\,0\,1\,0 \\ 0\,0\,\cdots\,0\,0\,0 \\ 0\quad\ 0\,0\,1 \\ \ddots\quad 0\,0 \\ \vdots\ \vdots \\ 0\,0 \\ 1 \end{pmatrix}$$

and the result agrees with [170, Corollary 5.5]

It is well-know that local cohomology modules as well as free resolutions depend on the characteristic of the base field, the most recurrent example being the Stanley–Reisner ideal associated to a minimal triangulation of $\mathbb{P}^2_{\mathbb{R}}$. Thus, Lyubeznik numbers also depend on the characteristic.

Example 56. Consider the ideal in $R = k[x_1, \ldots, x_6]$:

$$I = (x_1x_2x_3,\ x_1x_2x_4,\ x_1x_3x_5,\ x_2x_4x_5,\ x_3x_4x_5,$$

$$x_2x_3x_6,\ x_1x_4x_6,\ x_3x_4x_6,\ x_1x_5x_6,\ x_2x_5x_6)$$

The Lyubeznik table in characteristic zero and two are respectively:

$$\Lambda_{\mathbb{Q}}(R/I) = \begin{pmatrix} 0\,0\,0\,0 \\ 0\,0\,0 \\ 0\,0 \\ 1 \end{pmatrix}, \quad \Lambda_{\mathbb{Z}/2\mathbb{Z}}(R/I) = \begin{pmatrix} 0\,0\,1\,0 \\ 0\,0\,0 \\ 0\,1 \\ 1 \end{pmatrix}.$$

3.4 Injective Resolution of Local Cohomology Modules

The methods developed in the previous section allow us to describe the Bass numbers in the minimal $\mathbb{Z}^n$-graded injective resolution of a module with variation zero M. That is:

$$\mathbb{I}^\bullet(M): \quad 0 \longrightarrow I^0 \xrightarrow{\ d^0\ } I^1 \xrightarrow{\ d^1\ } \cdots \xrightarrow{\ d^{m-1}\ } I^m \xrightarrow{\ d^m\ } \cdots,$$

where the j-th term is

$$I^j = \bigoplus_{\alpha \in \{0,1\}^n} E_\alpha^{\mu_j(\mathfrak{p}_\alpha, M)}.$$

In particular we are able to compute the $\mathbb{Z}^n$-graded injective dimension of M or the $\mathbb{Z}^n$-graded small support. In this section we will take a close look to the structure of this minimal injective resolution and we will give a bound for the injective dimension.

3.5   Structure of the Injective Resolution

Let $(R, \mathfrak{m}, k)$ be a local ring and let M be an R-module. Bass numbers of finitely generated modules are known to satisfy the following properties:

(1) $\mu_i(\mathfrak{p}, M) < +\infty$, $\forall i$, $\forall \mathfrak{p} \in \mathrm{Supp}_R M$.
(2) Let $\mathfrak{p} \subseteq \mathfrak{q} \in \mathrm{Spec}\, R$ such that $\mathrm{ht}\,(\mathfrak{q}/\mathfrak{p}) = s$. Then

$$\mu_i(\mathfrak{p}, M) \neq 0 \Longrightarrow \mu_{i+s}(\mathfrak{q}, M) \neq 0.$$

(3) $\mathrm{id}_R M := \sup\{i \in \mathbb{Z} \mid \mu_i(\mathfrak{m}, M) \neq 0\}$.
(4) $\mathrm{depth}_R M \leq \dim_R M \leq \mathrm{id}_R M$.

When M is not finitely generated, similar properties for Bass numbers are known for some special cases. A.M. Simon [176] proved that properties (2) and (3) are still true for complete modules and M. Hellus [87] proved that $\dim_R M \leq \mathrm{id}_R M$ for cofinite modules.

For the case of local cohomology modules, C. Huneke and R. Sharp [110] and G. Lyubeznik [132, 133], proved that for a regular local ring $(R, \mathfrak{m}, k)$ containing a field k:

(1) $\mu_i(\mathfrak{p}, H_I^r(R)) < +\infty$, $\forall i$, $\forall r$, $\forall \mathfrak{p} \in \mathrm{Supp}_R H_I^r(R)$.
(4') $\mathrm{id}_R H_I^r(R) \leq \dim_R \mathrm{Supp}_R H_I^r(R)$.

In this section we want to study property (2) for the particular case of local cohomology modules supported on monomial ideals and give a sharper bound to (4') in terms of the small support.

We start with the following well-known general result on the minimal primes in the support of local cohomology modules.

Proposition 57. *Let $(R, \mathfrak{m})$ be a regular local ring containing a field k, $I \subseteq R$ be any ideal and $\mathfrak{p} \in \mathrm{Supp}_R H_I^r(R)$ be a minimal prime. Then we have $\mu_0(\mathfrak{p}, H_I^r(R)) \neq 0$, $\mu_i(\mathfrak{p}, H_I^r(R)) = 0\ \forall i > 0$.*

Corollary 58. *Let $(R, \mathfrak{m}, k)$ be a regular local ring containing a field k and $I \subseteq R$ be any ideal. If $\mathfrak{p} \in \mathrm{Supp}_R H_I^r(R)$ is minimal then $\mathfrak{p} \in \mathrm{supp}_R H_I^r(R)$. Thus, $\mathrm{Supp}_R H_I^r(R)$ and $\mathrm{supp}_R H_I^r(R)$ have the same minimal primes.*

If we had an analogue of property (2) then it would be enough to check those primes having 0-th Bass number different from zero, i.e. associated primes of $H_I^r(R)$, in order to get a first approach to the structure of its injective resolution. This is not the case as we can see in the following examples.

Example 59. Consider the ideal $I = (x_1, x_2, x_5) \cap (x_3, x_4, x_5) \cap (x_1, x_2, x_3, x_4)$. The non-vanishing pieces of the hypercube associated to the corresponding local cohomology modules are:

$$[H_I^3(R)]_\alpha = k \quad \text{for } \alpha = (1, 1, 0, 0, 1), (0, 0, 1, 1, 1).$$

$$[H_I^4(R)]_\alpha = k \quad \text{for } \alpha = (1, 1, 1, 1, 0), (1, 1, 1, 1, 1).$$

The Bass numbers of $H_I^3(R)$ and $H_I^4(R)$ are respectively

$\mathfrak{p}_\alpha$	μ_0	μ_1	μ_2
(x_1, x_2, x_5)	1	–	–
(x_3, x_4, x_5)	1	–	–
(x_1, x_2, x_i, x_5)	–	1	–
(x_i, x_3, x_4, x_5)	–	1	–
$(x_1, x_2, x_3, x_4, x_5)$	–	–	2

$\mathfrak{p}_\alpha$	μ_0	μ_1	μ_2
(x_1, x_2, x_3, x_4)	1	–	–
$(x_1, x_2, x_3, x_4, x_5)$	1	–	–

Notice that $\mathfrak{m} = (x_1, x_2, x_3, x_4, x_5)$ is not a minimal prime in the support of $H_I^4(R)$ but $\mu_0(\mathfrak{m}, H_I^4(R)) \neq 0$, $\mu_i(\mathfrak{m}, H_I^4(R)) = 0 \; \forall i > 0$.

Example 60. Consider the ideal $I = (x_1, x_4) \cap (x_2, x_5) \cap (x_1, x_2, x_3)$. The non-vanishing pieces of the hypercube associated to the corresponding local cohomology modules are:

$$[H_I^2(R)]_\alpha = k \quad \text{for } \alpha = (1, 0, 0, 1, 0), (0, 1, 0, 0, 1).$$

$$[H_I^3(R)]_\alpha = k \quad \text{for } \alpha = (1, 1, 1, 0, 0), (1, 1, 1, 1, 0), (1, 1, 1, 0, 1), (1, 1, 0, 1, 1),$$
$$(1, 1, 1, 1, 1).$$

The Bass numbers of $H_I^2(R)$ and $H_I^3(R)$ are respectively

$\mathfrak{p}_\alpha$	μ_0	μ_1	μ_2	μ_3
(x_1, x_4)	1	–	–	–
(x_2, x_5)	1	–	–	–
(x_1, x_4, x_i)	–	1	–	–
(x_2, x_5, x_i)	–	1	–	–
(x_1, x_4, x_i, x_j)	–	–	1	–
(x_2, x_5, x_i, x_j)	–	–	1	–
$(x_1, x_2, x_3, x_4, x_5)$	–	–	–	2

$\mathfrak{p}_\alpha$	μ_0	μ_1	μ_2
(x_1, x_2, x_3)	1	–	–
(x_1, x_2, x_3, x_4)	–	–	–
(x_1, x_2, x_3, x_5)	–	–	–
(x_1, x_2, x_4, x_5)	1	–	–
$(x_1, x_2, x_3, x_4, x_5)$	–	1	–

Notice that the small support and the support of $H_I^3(R)$ have the same minimal primes but (x_1, x_2, x_3, x_4) and (x_1, x_2, x_3, x_5) do not belong to $\mathrm{supp}_R(H_I^3(R))$.

From now on we will stick to the case of local cohomology modules supported on squarefree monomial ideals. We are going to consider chains of prime face ideals

$$\mathfrak{p}_0 \subseteq \mathfrak{p}_1 \subseteq \cdots \subseteq \mathfrak{m}$$

in the support of $M = H_I^r(R)$ such that $\mathfrak{p}_0$ is minimal. The Bass numbers with respect to $\mathfrak{p}_0$ are completely determined and, even though property (2) is no longer true, we will see that we have some control on the Bass numbers of $\mathfrak{p}_i$ depending on the structure of the corresponding n-hypercube. For simplicity, we will assume that $\mathfrak{p}_i$ is a face ideal $\mathfrak{p}_\alpha \subseteq \mathfrak{m}$ of height $n-1$ and $x_n \in \mathfrak{m} \setminus \mathfrak{p}_\alpha$ and that the Bass numbers with respect to $\mathfrak{p}_\alpha$ are known.

Let $\mathcal{M}^\bullet$ be the complex associated to the n-hypercube of M. For any $\beta \in \{0,1\}^n$, let $\mathcal{M}^\bullet_{\leq \beta}$ (resp. $\mathcal{M}^\bullet_{\geq \beta}$) be the subcomplex of $\mathcal{M}^\bullet$ with pieces of degree $\leq \beta$ (resp. $\geq \beta$). We have the short exact sequence of complexes

$$0 \longleftarrow \mathcal{M}^\bullet_{\leq \alpha} \longleftarrow \mathcal{M}^\bullet \longleftarrow \mathcal{M}^\bullet_{\geq 1-\alpha} \longleftarrow 0$$

Example 61. The short exact sequence

$$0 \longleftarrow \mathcal{M}^\bullet_{\leq(1,1,0)} \longleftarrow \mathcal{M}^\bullet \longleftarrow \mathcal{M}^\bullet_{\geq(0,0,1)} \longleftarrow 0$$

can be visualized from the corresponding 3-hypercube as follows:

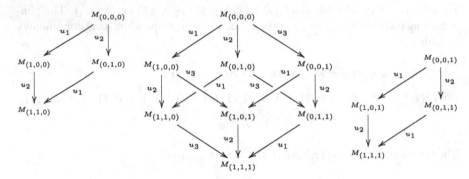

At this point we should notice the following key observations:

(a) The complex $\mathcal{M}^\bullet_{\leq \alpha}$ (resp. $\mathcal{M}^\bullet$) allows us to compute the Bass numbers with respect to $\mathfrak{p}_\alpha$ (resp $\mathfrak{m}$), i.e.

$$\mu_p(\mathfrak{p}_\alpha, M) = \dim_k H_p(\mathcal{M}^\bullet_{\leq \alpha}).$$

$$\mu_p(\mathfrak{m}, M) = \dim_k H_p(\mathcal{M}^\bullet).$$

(b) Consider the long exact sequence associated to the short exact sequence of complexes

$$\cdots \longleftarrow H_p(\mathcal{M}^\bullet) \longleftarrow H_p(\mathcal{M}^\bullet_{\geq 1-\alpha}) \overset{\delta^p}{\longleftarrow} H_p(\mathcal{M}^\bullet_{\leq \alpha}) \longleftarrow H_{p+1}(\mathcal{M}^\bullet) \longleftarrow \cdots$$

It might be useful to view it as

$$\cdots \longleftarrow [H_{\mathfrak{m}}^{p}(M)]_1 \longleftarrow [H_{\mathfrak{p}_\alpha}^{p}(M)]_1 \overset{\delta^p}{\longleftarrow} [H_{\mathfrak{p}_\alpha}^{p}(M)]_\alpha \longleftarrow [H_{\mathfrak{m}}^{p+1}(M)]_1 \longleftarrow \cdots$$

or even as the complex

$$\cdots \longleftarrow k^{\mu_p(\mathfrak{m},M)} \longleftarrow [H_{\mathfrak{p}_\alpha}^{p}(M)]_1 \overset{\delta^p}{\longleftarrow} k^{\mu_p(\mathfrak{p}_\alpha,M)} \longleftarrow k^{\mu_{p+1}(\mathfrak{m},M)} \longleftarrow \cdots$$

The connecting morphisms δ^p are the classes, in the corresponding homology groups, of the canonical morphisms u_n that describe the n-hypercube of M.

(c) The "difference" between $\mu_p(\mathfrak{p}_\alpha, M)$ and $\mu_{p+1}(\mathfrak{m}, M)$, i.e. the "difference" between $H_p(\mathcal{M}_{\leq\alpha}^\bullet)$ and $H_{p+1}(\mathcal{M}^\bullet)$, comes from the homology of the complex $\mathcal{M}_{\geq 1-\alpha}^\bullet$. Roughly speaking, it comes from the contribution of other chains of prime face ideals $\mathfrak{q}_0 \subseteq \mathfrak{q}_1 \subseteq \cdots \subseteq \mathfrak{m}$ in the support of a local cohomology module $M = H_I^r(R)$ such that $\mathfrak{q}_0$ is minimal and not containing $\mathfrak{p}_\alpha$.

In the sequel we will discuss how these connecting morphisms δ^p, i.e. these canonical maps u_n control the contribution of the Bass numbers with respect to $\mathfrak{p}_\alpha$ to the Bass numbers with respect to $\mathfrak{m}$.

Discussion 1. Consider the case where $\mathfrak{p}_\alpha \subseteq \mathfrak{m}$ is a minimal prime of height $n-1$ in the support of M. We have $\mu_0(\mathfrak{p}_\alpha, M) = 1$ and $\mu_p(\mathfrak{p}_\alpha, M) = 0$ for all $p > 0$. Then, the long exact sequence is just

$$0 \longleftarrow [H_{\mathfrak{m}}^{0}(M)]_1 \longleftarrow [H_{\mathfrak{p}_\alpha}^{0}(M)]_1 \overset{\delta^0}{\longleftarrow} k \longleftarrow [H_{\mathfrak{m}}^{1}(M)]_1 \longleftarrow [H_{\mathfrak{p}_\alpha}^{1}(M)]_1 \longleftarrow 0$$

and $[H_{\mathfrak{m}}^{i}(M)]_1 \cong [H_{\mathfrak{p}_\alpha}^{i}(M)]_1$, for all $i \geq 2$, so $\mathfrak{p}_\alpha$ contributes to $\mu_p(\mathfrak{m}, M)$ for $p = 0, 1$. In particular, we have:

- $\mu_0(\mathfrak{m}, M) = 0$ if and only if $[H_{\mathfrak{p}_\alpha}^{0}(M)]_1 = 0$ or $[H_{\mathfrak{p}_\alpha}^{0}(M)]_1 = k$ and $\delta^0 \neq 0$.
- $\mu_1(\mathfrak{m}, M) = 0$ if and only if $[H_{\mathfrak{p}_\alpha}^{1}(M)]_1 = 0$ and $\delta^0 \neq 0$.

Discussion 2. In general, let $s = \max\{i \in \mathbb{Z}_{\geq 0} \mid \mu_i(\mathfrak{p}_\alpha, M) \neq 0\}$, then we have

$$\cdots \longleftarrow [H_{\mathfrak{p}_\alpha}^{s}(M)]_1 \overset{\delta^s}{\longleftarrow} k^{\mu_s(\mathfrak{p}_\alpha,M)} \longleftarrow [H_{\mathfrak{m}}^{s+1}(M)]_1 \longleftarrow [H_{\mathfrak{p}_\alpha}^{s+1}(M)]_1 \longleftarrow 0$$

and $[H_{\mathfrak{m}}^{i}(M)]_1 \cong [H_{\mathfrak{p}_\alpha}^{i}(M)]_1$, for all $i \geq s+2$, so $\mathfrak{p}_\alpha$ contributes to $\mu_p(\mathfrak{m}, M)$ for $p \leq s+1$. Again, we can describe conditions for the vanishing of $\mu_s(\mathfrak{m}, M)$ and $\mu_{s+1}(\mathfrak{m}, M)$ in terms of the connecting morphism δ^s. One can find examples where any situation is possible, i.e.

- $\mu_s(\mathfrak{p}_\alpha, M) \neq 0$ and $\mu_s(\mathfrak{m}, M) = 0$, $\mu_{s+1}(\mathfrak{m}, M) \neq 0$.
- $\mu_s(\mathfrak{p}_\alpha, M) \neq 0$ and $\mu_s(\mathfrak{m}, M) \neq 0$, $\mu_{s+1}(\mathfrak{m}, M) = 0$.
- $\mu_s(\mathfrak{p}_\alpha, M) \neq 0$ and $\mu_s(\mathfrak{m}, M) \neq 0$, $\mu_{s+1}(\mathfrak{m}, M) \neq 0$.
- $\mu_s(\mathfrak{p}_\alpha, M) \neq 0$ and $\mu_s(\mathfrak{m}, M) = 0$, $\mu_{s+1}(\mathfrak{m}, M) = 0$.

Discussion 3. In the case that there exists a prime ideal $\mathfrak{p}_\alpha \notin \text{supp}_R M$, then we have $[H_{\mathfrak{m}}^i(M)]_1 \cong [H_{\mathfrak{p}_\alpha}^i(M)]_1$, for all i. Therefore, the contribution to the Bass numbers $\mu_p(\mathfrak{m}, M)$ comes from other chains of prime face ideals $\mathfrak{q}_0 \subseteq \mathfrak{q}_1 \subseteq \cdots \subseteq \mathfrak{m}$.

Remark 62. Consider the largest chain of face ideals $\mathfrak{p}_0 \subseteq \mathfrak{p}_1 \subseteq \cdots \subseteq \mathfrak{p}_n$ in the small support of a local cohomology module $H_I^r(R)$. In this best case scenario we have a version of property (2) that we introduced at the beginning of this section that reads off as:

- $\mu_0(\mathfrak{p}_0, H_I^r(R)) = 1$ and $\mu_j(\mathfrak{p}_0, H_I^r(R)) = 0 \; \forall j > 0$.
- $\mu_i(\mathfrak{p}_i, H_I^r(R)) \neq 0$ and $\mu_j(\mathfrak{p}_i, H_I^r(R)) = 0 \; \forall j > i$, for all $i = 1, \ldots, n$.

For example, when R/I is Cohen–Macaulay, we have

$$\mu_p(\mathfrak{p}_\alpha, H_I^{\text{ht}\, I}(R)) = \begin{cases} 0 & p \neq n - |\alpha| \\ 1 & p = n - |\alpha| \end{cases}$$

for all face ideals in the support of R/I, so the injective resolution of $H_I^{\text{ht}\, I}(R)$ behaves like the injective resolution of a Gorenstein ring. In particular, its injective resolution is linear.

On the other end of possible cases we may have:

- $\mu_0(\mathfrak{p}_0, H_I^r(R)) = \mu_0(\mathfrak{p}_n, H_I^r(R)) = 1$ and
- $\mu_j(\mathfrak{p}_0, H_I^r(R)) = \mu_j(\mathfrak{p}_n, H_I^r(R)) = 0 \; \forall j > 0$.

Notice that in this case the same property holds for any prime ideal $\mathfrak{p}_i$ in the chain. In particular all the primes in the chain are associated primes of $H_I^r(R)$.

3.6 Injective Dimension

It follows from the previous discussions that the length of the injective resolution of the local cohomology module $H_I^r(R)$ has a controlled growth when we consider chains of prime face ideals $\mathfrak{p}_0 \subseteq \mathfrak{p}_1 \subseteq \cdots \subseteq \mathfrak{m}$ starting with a minimal prime ideal $\mathfrak{p}_0$.

Proposition 63. *Let $I \subseteq R = k[x_1, \ldots, x_n]$ be a squarefree monomial ideal and set*

$$s := \max\{i \in \mathbb{Z}_{\geq 0} \mid \mu_i(\mathfrak{p}_\alpha, H_I^r(R)) \neq 0\}$$

for all prime ideals $\mathfrak{p}_\alpha \in \text{Supp}_R H_I^r(R)$ such that $|\alpha| = n - 1$.
Then $\mu_t(\mathfrak{m}, H_I^r(R)) = 0 \; \forall t > s + 1$.

Therefore we get the main result of this section:

Theorem 64. *Let $I \subseteq R = k[x_1, \ldots, x_n]$ be a squarefree monomial ideal. Then, $\forall r$ we have*

$$^* \mathrm{id}_R H_I^r(R) \leq \dim_R{}^* \mathrm{supp}_R H_I^r(R).$$

Using [78, Theorem 1.2.3] we also have

$$\mathrm{id}_R H_I^r(R) \leq \dim_R \mathrm{supp}_R H_I^r(R).$$

The small support might be strictly included in the support so we get a sharper bound for the injective dimension of local cohomology modules supported on monomial ideals, i.e. the supremum of the lengths r, taken over all strict chains $\mathfrak{p}_0 \subset \mathfrak{p}_1 \subset \cdots \subset \mathfrak{p}_r$ of prime ideals in the small support. We have

$$\mathrm{id}_R H_I^r(R) \leq \dim_R \mathrm{supp}_R H_I^r(R) \leq \dim_R \mathrm{Supp}_R H_I^r(R)$$

and we may find examples where the dimension of the small support is strictly smaller than the dimension of the support.

Appendix: $\mathbb{Z}^n$-Graded Free and Injective Resolutions

The theory of $\mathbb{Z}^n$-graded rings and modules is analogous to that of $\mathbb{Z}$-graded rings and modules. The aim of this appendix is to fix the notation that we use throughout this work. For a detailed exposition of these topics we refer to [30, 78, 141].

In this work, we consider the case of modules over the polynomial ring $R = k[x_1, \ldots, x_n]$, where k is a field of any characteristic and $x_1, \ldots, x_n$ are independent variables. Let $\varepsilon_1, \ldots, \varepsilon_n$ be the canonical basis of $\mathbb{Z}^n$. Then, the ring R has a natural $\mathbb{Z}^n$-graduation given by $\deg(x_i) = \varepsilon_i$. Henceforth, the term graded will always mean $\mathbb{Z}^n$-graded. If $M = \bigoplus_{\alpha \in \mathbb{Z}^n} M_\alpha$ is a graded R-module and $\beta \in \mathbb{Z}^n$, as usual we denote by $M(\beta)$ the shifted graded R-module whose underlying R-module structure is the same as that of M and where the grading is given by $(M(\beta))_\alpha = M_{\beta + \alpha}$. In particular, the free R-module of rank one generated in degree $\alpha \in \mathbb{Z}^n$ is $R(-\alpha)$.

The category of $\mathbb{Z}^n$-graded modules will be denoted $^* \mathrm{Mod}(R)$.

3.7 Free Resolutions

The minimal graded free resolution of a monomial ideal $J \subseteq R$ is an exact sequence of free $\mathbb{Z}^n$-graded R-modules:

$$\mathbb{L}_\bullet(J): \quad 0 \longrightarrow L_m \xrightarrow{d_m} \cdots \longrightarrow L_1 \xrightarrow{d_1} L_0 \longrightarrow J \longrightarrow 0$$

where the j-th term is of the form

$$L_j = \bigoplus_{\alpha \in \mathbb{Z}^n} R(-\alpha)^{\beta_{j,\alpha}(J)},$$

and the matrices of the morphisms $d_j : L_j \longrightarrow L_{j-1}$ do not contain invertible elements. From this expression we can get the following:

- The **projective dimension** of J, denoted $\mathrm{pd}(J)$, is the greatest homological degree in the resolution. Namely

$$\mathrm{pd}(J) := \max\{j \mid L_j \neq 0\}.$$

- The $\mathbb{Z}^n$-graded **Betti numbers** of J are the invariants defined by $\beta_{j,\alpha}(J)$. Betti numbers can also be described as:

$$\beta_{j,\alpha}(J) = \dim_k \mathrm{Tor}_j^R(J, k)_\alpha.$$

- The **Castelnuovo–Mumford regularity** of J denoted $\mathrm{reg}(J)$ is

$$\mathrm{reg}(J) := \max\{|\alpha| - j \mid \beta_{j,\alpha}(J) \neq 0\}.$$

- Given an integer r, the r-**linear strand** of $\mathbb{L}_\bullet(J)$ is the complex:

$$\mathbb{L}_\bullet^{<r>}(J): \quad 0 \longrightarrow L_m^{<r>} \xrightarrow{d_m^{<r>}} \cdots \longrightarrow L_1^{<r>} \xrightarrow{d_1^{<r>}} L_0^{<r>} \longrightarrow 0 \;,$$

where

$$L_j^{<r>} = \bigoplus_{|\alpha|=j+r} R(-\alpha)^{\beta_{j,\alpha}(J)},$$

and the differentials $d_j^{<r>} : L_j^{<r>} \longrightarrow L_{j-1}^{<r>}$ are the corresponding components of d_j.

The study of Betti numbers of any monomial ideal can be reduced to the understanding of the squarefree monomial case but, even in this case, describing the minimal free resolution may be hard. This has been a very active area of research so we can not cover all the existent literature in this survey. We only want to point out that, apart from describing the minimal free resolution of particular cases of monomial ideals, some successful lines of research are:

- Construct possibly non-minimal free resolutions. The most famous one being:

  - *Taylor resolution [188]:* Let $\{\mathbf{x}^{\alpha_1}, \ldots, \mathbf{x}^{\alpha_m}\}$ be a set of generators of I. Let L be the free R-module of rank m generated by $e_1, \ldots, e_m$. The Taylor complex $\mathbb{T}_\bullet(R/I)$ is of the form:

$$\mathbb{T}_{\bullet}(R/I) : \quad 0 \longrightarrow L_m \xrightarrow{d_m} \cdots \longrightarrow L_1 \xrightarrow{d_1} L_0 \longrightarrow 0 ,$$

where $L_j = \bigwedge^j L$ is the j-th exterior power of L and if lcm denotes the least common multiple, then the differentials d_j are defined as

$$d_j (e_{i_1} \wedge \cdots \wedge e_{i_j})$$
$$= \sum_{1 \le k \le j} (-1)^k \frac{\operatorname{lcm} (\mathbf{x}^{\alpha_{i_1}}, \ldots, \mathbf{x}^{\alpha_{i_j}})}{\operatorname{lcm} (\mathbf{x}^{\alpha_{i_1}}, \ldots, \widehat{\mathbf{x}^{\alpha_{i_k}}}, \ldots, \mathbf{x}^{\alpha_{i_j}})} \; e_{i_1} \wedge \cdots \wedge \widehat{e_{i_k}} \wedge \cdots \wedge e_{i_j}.$$

Some subcomplexes of the Taylor resolution as the *Lyubeznik resolution* [130] or the *Scarf complex* [16] also have a prominent role in the theory. These are examples of *simplicial resolutions*. The idea behind the concept of simplicial (resp. cellular, CW) resolution introduced in [16] (see also [17, 116]) is to associate to a free resolution of an ideal a simplicial complex (resp. regular cell complex, CW-complex) that carries in its structure the algebraic structure of the free resolution.

- Relating the $\mathbb{Z}^n$-graded Betti numbers of monomial ideals to combinatorial objects described by the generators of the monomial ideal. Here we highlight:

  - *Hochster's formula:* Via the Stanley–Reisner correspondence we associate a simplicial complex Δ to our squarefree monomial ideal $I = I_\Delta$. Then, for a given face $\sigma_\alpha \in \Delta$ consider the restriction $\Delta_\alpha := \{ \tau \in \Delta \mid \tau \in \sigma_\alpha \}$. Then we have

    $$\beta_{i,\alpha}(I_\Delta) = \dim_k \tilde{H}_{|\alpha|-i-2}(\Delta_\alpha; k)$$

    where $\tilde{H}(-; k)$ denotes reduced simplicial homology. Here we agree that the reduced homology with coefficients in k of the empty simplicial complex is k in degree -1 and zero otherwise.
  - *LCM-lattice [75]:* The lcm-lattice of a monomial ideal is the poset $P(I)$ of the least common multiples of the generators of the ideal. To any poset one can associate the *order complex*, this is a simplicial complex which has as vertices the elements of the poset and where a set of vertices $p_0, \ldots, p_r$ determines a r-dimensional simplex if $p_0 < \cdots < p_r$. We define $K(> p)$ to be the simplicial complex attached to the subposet $\{ q \in P(I) \mid q > p \}$. Then,

    $$\beta_{i,p}(I_\Delta) = \dim_k \tilde{H}_{i-2}(K(> p); k).$$

    It follows that if two ideals have isomorphic lcm-lattices then they have the same Betti numbers up to some relabeling of the degrees.

3.8 Injective Resolutions

If M is a $\mathbb{Z}^n$-graded module one can define its $\mathbb{Z}^n$-graded injective envelope $^* E(M)$. Therefore, the category $^*\mathrm{Mod}(R)$ of $\mathbb{Z}^n$-graded modules has enough injectives and a $\mathbb{Z}^n$-graded version of the Matlis–Gabriel theorem holds: the indecomposable injective objects of $^*\mathrm{Mod}(R)$ are the shifted injective envelopes $^* E(R/\mathfrak{p}_\alpha)(\beta)$, where $\mathfrak{p}_\alpha$ is a face ideal of R and $\beta \in \mathbb{Z}^n$, and every graded injective module is isomorphic to a unique (up to order) direct sum of indecomposable injectives.

The minimal graded injective resolution of $\mathbb{Z}^n$-graded R-module M is a sequence:

$$\mathbb{I}^\bullet(M): \quad 0 \xrightarrow{} I^0 \xrightarrow{d^0} I^1 \xrightarrow{d^1} \cdots \xrightarrow{} I^m \xrightarrow{d^m} \cdots ,$$

exact everywhere except the 0-th step such that $M = \mathrm{Ker}\,(d^0)$, the j-th term is

$$I^j = \bigoplus_{\alpha \in \mathbb{Z}^n} {}^* E(R/\mathfrak{p}_\alpha)(\beta)^{\mu_j(\mathfrak{p}_\alpha, M)},$$

and I^j is the injective envelope of $\mathrm{Ker}\,d^j$. From this expression we can get the following:

- The $\mathbb{Z}^n$-**graded injective dimension** of M, denoted $^* \mathrm{id}_R(M)$, is the greatest cohomological degree in the minimal graded injective resolution. Namely

$$^* \mathrm{id}_R(M) = \max\{j \mid I^j \neq 0\}.$$

Remark 65. The notation $^*\mathrm{id}_R$ usually refers to the $\mathbb{Z}$-graded injective dimension but we use the same notation for the $\mathbb{Z}^n$-graded injective dimension as in [200]. The reader must be aware that both concepts are different but in this work no confusion is possible since we only consider the $\mathbb{Z}^n$-graded case.

- The **Bass numbers** of M are the invariants defined by $\mu_j(\mathfrak{p}_\alpha, M)$. By using the results of [78], these numbers are equal to the usual Bass numbers that appear in the minimal injective resolution of M. So, they can also be computed as

$$\mu_j(\mathfrak{p}_\alpha, M) = \dim_{k(\mathfrak{p}_\alpha)} \mathrm{Ext}_R^j(k(\mathfrak{p}_\alpha), M_{\mathfrak{p}_\alpha}).$$

If we want to compute the Bass numbers with respect to any prime ideal we have to refer to the result of S. Goto and K.I. Watanabe [78, Theorem 1.2.3]. Namely, given any prime ideal $\mathfrak{p} \in \mathrm{Spec}\,R$, let $\mathfrak{p}_\alpha$ be the largest face ideal contained in $\mathfrak{p}$. If $\mathrm{ht}\,(\mathfrak{p}/\mathfrak{p}_\alpha) = s$ then $\mu_p(\mathfrak{p}_\alpha, M) = \mu_{p+s}(\mathfrak{p}, M)$. Notice that in general we have $^*\mathrm{id}_R M \leq \mathrm{id}_R M$.

- The **small support** of M introduced by H.B. Foxby [63] is defined as

$$\mathrm{supp}_R M := \{\mathfrak{p} \in \mathrm{Spec}\, R \mid \mathrm{depth}_{R_\mathfrak{p}} M_\mathfrak{p} < \infty\},$$

where $\mathrm{depth}_R M := \inf\{i \in \mathbb{Z} \mid \mathrm{Ext}_R^i(R/\mathfrak{m}, M) \neq 0\} = \inf\{i \in \mathbb{Z} \mid \mu_i(\mathfrak{m}, M) \neq 0\}$. In terms of Bass numbers we have that $\mathfrak{p} \in \mathrm{supp}_R M$ if and only if there exists some integer $i \geq 0$ such that $\mu_i(\mathfrak{p}, M) \neq 0$. It is also worth to point out that $\mathrm{supp}_R M \subseteq \mathrm{Supp}_R M$, and equality holds when M is finitely generated. It follows that in general we have $\dim_R \mathrm{supp}_R M \leq \dim_R \mathrm{Supp}_R M$.

We can also define the $\mathbb{Z}^n$-graded small support that we denote $^*\mathrm{supp}_R M$ as the set of face ideals in the support of M that at least have a Bass number different from zero.

- Given an integer r, the r-**linear strand** of $\mathbb{I}_\bullet(M)$ is the complex:

$$\mathbb{I}_\bullet^{<r>}(M): \quad 0 \longrightarrow I_0^{<r>} \longrightarrow I_1^{<r>} \longrightarrow \cdots \longrightarrow I_m^{<r>} \longrightarrow 0 \ ,$$

where

$$I_j^{<r>} = \bigoplus_{|\alpha|=j+r} {}^*E(R/\mathfrak{p}_\alpha)(\beta)^{\mu_j(\mathfrak{p}_\alpha, M)}.$$

3.9 Monomial Matrices

The concept of *monomial matrices* was coined by E. Miller in [140] (see also [141]) to deal with maps of $\mathbb{Z}^n$-graded free, injective or flat modules in an unified way.

A matrix whose (p, q)-entry is of the form $\lambda_{pq} \mathbf{x}^{\beta_{pq}}$, where $\beta_{pq} \in \mathbb{Z}^n$, defines a map of $\mathbb{Z}^n$-graded free modules but we also have to keep track of the degrees of the generators in the source and the target of the matrix to determine the map uniquely. Once we do this, we can simplify the notation just using the scalars λ_{pq} as entries of our matrix

$$\begin{array}{c} \\ \beta_{1\cdot} \\ \vdots \\ \beta_{p\cdot} \\ \vdots \end{array} \begin{array}{ccc} \beta_{\cdot 1} & \cdots & \beta_{\cdot q} & \cdots \end{array} \\ \left(\begin{array}{cccc} & & & \\ & & & \\ & & \lambda_{pq} & \\ & & & \end{array} \right).$$

The nice idea in [140] is that, modifying conveniently the notation for the source and the target, one can use the same kind of matrices to describe maps between injective modules or flat modules. Now we allow $\beta_{pq} \in (\mathbb{Z} \cup *)^n$ where $*$ behaves like $-\infty$ except for $-1 \cdot * = *$.

There are several technical questions to be addressed for these monomial matrices to work out but we will skip the details. In this work we only have to have in mind that the matrices that describe a free resolution or a injective resolution have scalar entries with the appropriate source and target labels. This will be also true for the maps describing the Čech complex.

Acknowledgements Many thanks go to Anna M. Bigatti, Philippe Gimenez, and Eduardo Sáenz-de-Cabezón for the invitation to participate in the MONICA conference and the great enviroment they created there. I am also indebted with Oscar Fernández Ramos who agreed to develop the Macaulay 2 routines that not only allowed us to perform many computations but also enlightened part of my research.

Local Cohomology Using Macaulay2

Josep Àlvarez Montaner and Oscar Fernández-Ramos

Over the last 20 years there were many advances made in the computational theory of D-modules. Nowadays, the most common computer algebra systems[1] such as Macaulay2 or Singular have important available packages for working with D-modules. In particular, the package D-modules [127] for Macaulay 2 [80] developed by A. Leykin and H. Tsai contains an implementation of the algorithms given by U. Walther [194] and T. Oaku and N. Takayama [153] to compute local cohomology modules. For a quick overview of this general approach we recommend the interested reader to take a look at [126, 195]. We have to point out that the algorithm to compute Lyubeznik numbers given by U. Walther [194] is not available in this package (see [9] for an alternative approach).

The complexity of the algorithm turns out to be a major drawback when trying to compute even some basic examples such as local cohomology modules supported on monomial ideals. It should be noticed that F. Barkats [15] gave an algorithm to compute a presentation of this kind of modules. However she was only able to compute effectively examples in the polynomial ring $k[x_1, \ldots, x_6]$.

Josep Àlvarez Montaner was partially supported by SGR2009-1284 and MTM2010-20279-C02-01.

Oscar Fernández-Ramos was partially supported by MTM2010-20279-C02-02.

[1] CoCoA is still working on that.

J. Àlvarez Montaner (✉)
Departamento de Matemàtica Aplicada I, Universitat Politècnica de Catalunya, Av. Diagonal 647, Barcelona 08028, Spain
e-mail: Josep.Alvarez@upc.edu

O. Fernández-Ramos
Dipartimento di Matematica, Università degli Studi di Genova, Via Dodecaneso 35, 16146 Genova, Italy
e-mail: fernandez@dima.unige.it

A.M. Bigatti et al. (eds.), *Monomial Ideals, Computations and Applications*, Lecture Notes in Mathematics 2083, DOI 10.1007/978-3-642-38742-5_6, © Springer-Verlag Berlin Heidelberg 2013

In order to prepare the Tutorials of the course given at "MONICA: MONomial Ideals, Computations and Applications" we implemented our own functions to compute the characteristic cycle of local cohomology modules supported on monomial ideals and Lyubeznik numbers as well. We finally decided to use Macaulay2 since it already provides some packages to deal with edge ideals and simplicial complexes. In fact, our functions call other functions from the SimplicialComplexes package written by S. Popescu, G. G. Smith and M. Stillman. Also useful in this context, in order to experiment with many examples, is the EdgeIdeals package written by C. Francisco, A. Hoefel and A. Van Tuyl. For more information on this last package we recommend to take a look at the notes provided by A. Van Tuyl in chapter "Edge Ideals Using Macaulay2" of this volume.

The following functions are not included in any version of Macaulay2. We collected them in the file LCfunctionsv4.m2 that we posted in

    http://monica.unirioja.es/conference/monica_program.html

- multCCLC: It computes the multiplicities and the components of the character-istic variety of a local cohomology module supported on a squarefree monomial ideal. The input is just the squarefree monomial ideal and it returns as output a hashtable whose keys correspond to the cohomology degrees and the values are lists of pairs. Each pair consists of the multiplicity and the corresponding component of the characteristic variety. Each component is expressed as the variety of the defining ideal. For a more visual output we collect the varieties by their corresponding height using the command fancyOut with the output of multCCLC as input.
- lyubeznikTable: It computes the Lyubeznik table of a squarefree monomial ideal. First we have to compute the minimal free resolution of its Alexander dual ideal. Then, the command linearStrands extracts the linear strands of this resolution and finally, the command lyubeznikTable computes the homology groups of the linear strands and displays the correponding Lyubeznik numbers in a BettiTally as output.

1 Getting Started

Before starting you will have to save the source code in your working directory. It can be obtained using the command path in a running session of Macaulay2:

```
Macaulay2, version 1.4
with packages: ConwayPolynomials, Elimination, IntegralClosure, LLLBases,
               PrimaryDecomposition, ReesAlgebra, TangentCone

i1 : path

o1 = {./, .Macaulay2/code/, .Macaulay2/local/share/Macaulay2/,
     ----------------------------------------------------------------
     .Macaulay2/local/common/share/Macaulay2/, /usr/share/Macaulay2/}

o1 : List
```

where ./ means the directory from where you run Macaulay2. You can also check it out using currentDirectory(). Once we have this package installed we are ready to start our session. First we will have to load it

```
i2 : load "LCfunctionsv4.m2"
```

Then we introduce our favorite polynomial ring, but we have to make sure that we give the appropriate $\mathbb{Z}^n$-grading to the variables.

```
i3 : R=QQ[x_1..x_5,DegreeRank=>5]

o3 = R

o3 : PolynomialRing

i4 : R_0

o4 = x
      1

o4 : R

i5 : degree R_0

o5 = {1, 0, 0, 0, 0}

o5 : List
```

Now we introduce a squarefree monomial ideal. Just for completeness we check out its minimal primary decomposition. We also compute its Alexander dual ideal since we will need it later on.

```
i6 : I=monomialIdeal (x_1*x_2,x_2*x_4,x_1*x_5,x_3*x_4*x_5)

o6 = monomialIdeal (x x , x x , x x , x x x )
                     1 2    2 4    1 5    3 4 5

o6 : MonomialIdeal of R

i7 : primaryDecomposition I

o7 = {monomialIdeal (x , x ), monomialIdeal (x , x ),
                      1   4                   2   5
     ----------------------------------------------------
     monomialIdeal (x , x , x )}
                     1   2   3

o7 : List

i8 : Idual=dual I

o8 = monomialIdeal (x x x , x x , x x )
                     1 2 3   1 4   2 5

o8 : MonomialIdeal of R
```

The characteristic cycle of the corresponding local cohomology modules are

```
i9 : multCCLC I

o9 = HashTable{2 => {{1, (x , x )}, {1, (x , x )}}                        }
                           1   4          2   5
               3 => {{1, (x , x , x )}, {1, (x , x , x , x )},
                          1   2   3          1   2   3   5
                     --------------------------------------------------
                     {1, (x , x , x , x )}, {1, (x , x , x , x )},
                          1   2   3   4          1   2   4   5
                     --------------------------------------------------
                     {1, (x , x , x , x , x )}}
                          1   2   3   4   5
               4 => {}
o9 : HashTable

i10 : fancyOut oo

o10 = HashTable{H^2 => {{(1, V(x_1,x_4))}}                                }
                       {(1, V(x_2,x_5))}
                H^3 => {{(1, V(x_1,x_2,x_3))}, {(1, V(x_1,x_2,x_3,x_5))},
                                               {(1, V(x_1,x_2,x_3,x_4))}
                                               {(1, V(x_1,x_2,x_4,x_5))}
                       --------------------------------------------------
                       {(1, V(x_1,x_2,x_3,x_4,x_5))}}}
o10 : HashTable
```

Notice that we obtain two local cohomology modules different from zero and we can easily describe the support of these modules and their dimension. We can also compare the multiplicities with the Betti numbers of the Alexander dual ideal.

```
i11 : r= res Idual

        1      3      3      1
o11 = R  <-- R  <-- R  <-- R  <-- 0

        0      1      2      3      4


o11 : ChainComplex

i12 : LS=linearStrands r

                  2
o12 = HashTable{2 => R                   }

                     1
                     1      3      1
                3 => R  <-- R  <-- R

                     1      2      3
                4 => 0

o12 : HashTable
```

```
i13 : B=betti r

            0 1 2 3
o13 = total: 1 3 3 1
          0: 1 . . .
          1: . 2 . .
          2: . 1 3 1

o13 : BettiTally
```

We already computed the linear strands of the minimal free resolution of the Alexander dual ideals so we are ready to compute the Lyubeznik table

```
i14 : T=lyubeznikTable LS

            0 1 2 3
o14 = total: . . 1 2
          0: . . . .
          1: . 1 . .
          2: . . . .
          3: . . . 2

o14 : BettiTally
```

2 Tutorial

Once we get acquainted with the use of these functions we propose several exercises. This is just a small sample of questions. We encourage the readers to experiment with different families of examples and come up with their own formulas for Lyubeznik numbers or, in general, Bass numbers. The question on how to find a general description of the injective resolution, i.e. Bass numbers and maps between injective modules, for any ideal might be too difficult. As in the case of free resolutions it would be interesting to study the different linear strands in the injective resolution of local cohomology modules.

A recurrent topic in recent years has been to attach a cellular structure to the free resolution of a monomial ideal. In general this can not be done as it is proved in [191] but there are large families of ideals having a cellular resolution. Using the dictionary we described in Sect. 2.9 of chapter "Local Cohomology Modules Supported on Monomial Ideals" we can translate the same questions to Lyubeznik numbers. In particular it would be interesting to find cellular structures on the linear strands of a free resolution so one can give a topological description of Lyubeznik numbers.

Exercise 1. Given the ideals

- $I_1 = (x_1 x_4, x_2 x_3)$,
- $I_2 = (x_1 x_2, x_1 x_3, x_2 x_4)$
- $I_3 = (x_1 x_2 x_3, x_1 x_4, x_2 x_4)$

a) Compute the characteristic cycle of the corresponding local cohomology modules.
b) Describe the support of these modules and compute its dimension.
c) Are the corresponding Stanley–Reisner rings Cohen–Macaulay, Gorenstein or Buchsbaum?

Exercise 2. Describe the n-hypercube of the local cohomology modules supported on the ideals of the previous exercise.

Exercise 3. Given the ideal

$$I = (x_1, x_2) \cap (x_3, x_4) \cap (x_5, x_6) \cap (x_7, x_8) \cap (x_9, x_1) \cap (x_9, x_2) \cap (x_9, x_3) \cap$$

$$\cap (x_9, x_4) \cap (x_9, x_5) \cap (x_9, x_6) \cap (x_9, x_7) \cap (x_9, x_8)$$

a) Compute the characteristic cycle of the corresponding local cohomology modules and construct the corresponding table $\Gamma(R/I)$.
b) Compute the Betti table of the Alexander dual ideal $I^\vee$.
c) Is I sequentially Cohen–Macaulay?
d) Compute its Lyubeznik table.

Hint: The EdgeIdeals package has the isSCM command to check out the sequentially Cohen–Macaulay property.

Exercise 4. Let $I = (x_1 x_2, x_2 x_3, x_3 x_4, x_4 x_1, x_1 x_5)$ be the edge ideal of a four-cycle with a whisker.

a) Is I sequentially Cohen–Macaulay?
b) Compute its Lyubeznik table.

Exercise 5. Consider your favorite sequentially Cohen–Macaulay ideal and compute its Lyubeznik table.

Exercise 6. Consider the ideal

$$I = (x_1, x_3) \cap (x_1, x_4) \cap (x_2, x_4) \cap (x_2, x_5) \cap (x_3, x_5).$$

a) Compute all the Bass numbers of the local cohomology module $H_I^2(R)$.
b) Is the injective resolution linear?

Exercise 7. Find examples of local cohomology modules $M = H_I^r(R)$ such that their Bass numbers satisfy

- $\mu_s(\mathfrak{p}_\alpha, M) \neq 0$ and $\mu_s(\mathfrak{m}, M) = 0$, $\mu_{s+1}(\mathfrak{m}, M) \neq 0$,
- $\mu_s(\mathfrak{p}_\alpha, M) \neq 0$ and $\mu_s(\mathfrak{m}, M) \neq 0$, $\mu_{s+1}(\mathfrak{m}, M) = 0$,
- $\mu_s(\mathfrak{p}_\alpha, M) \neq 0$ and $\mu_s(\mathfrak{m}, M) \neq 0$, $\mu_{s+1}(\mathfrak{m}, M) \neq 0$,
- $\mu_s(\mathfrak{p}_\alpha, M) \neq 0$ and $\mu_s(\mathfrak{m}, M) = 0$, $\mu_{s+1}(\mathfrak{m}, M) = 0$,

where $\mathfrak{p}_\alpha \subseteq \mathfrak{m}$ is a face ideal such that ht $(\mathfrak{m}/\mathfrak{p}_\alpha) = 1$.

Exercise 8. Compute all the Bass numbers of the following modules with variation zero in $R = k[x_1, \ldots, x_5]$:

- $M = H^4_{(x_1,x_2,x_3,x_4)}(R) \oplus E_{(0,0,1,1,1)}$.
- $M = H^3_I(R)$, where $I = (x_1x_4, x_2x_3, x_3x_4, x_1x_2x_5)$.

Are these modules isomorphic?

Exercise 9. Find a monomial ideal I such that

$$\mathrm{id}_R H^r_I(R) < \dim_R \mathrm{Supp}_R H^r_I(R)$$

for some r.

Exercise 10. Find a formula for the Lyubeznik table of the Alexander dual of the edge ideals $I(C_n)$ of the cycle graph $C_n, n \geq 3$.

Exercise 11. The same as in the previous exercise but for the complement C_n^c of the cycle graph $C_n, n \geq 3$.

Exercise 12. Compute the Lyubeznik table of the ideal associated to a minimal triangulation of $\mathbb{P}^2_{\mathbb{R}}$ when the characteristic of the field is 0 and 2:

$$I = (x_1x_2x_3, \ x_1x_2x_4, \ x_1x_3x_5, \ x_2x_4x_5, \ x_3x_4x_5,$$
$$x_2x_3x_6, \ x_1x_4x_6, \ x_3x_4x_6, \ x_1x_5x_6, \ x_2x_5x_6).$$

Exercise 13. Is there any ideal such that the local cohomology modules depend on the characteristic of the field but the Lyubeznik numbers do not?
Hint: Modify conveniently the ideal associated to a minimal triangulation of $\mathbb{P}^2_{\mathbb{R}}$.

References

1. J. Abbott, A.M. Bigatti, CoCoALib: A C++ library for doing Computations in Commutative Algebra (2011). Available at http://cocoa.dima.unige.it/cocoalib
2. A. Alilooee, S. Faridi, Betti numbers of path ideals of cycles and line (2011). Preprint [arXiv: 1110.6653v1]
3. J. Àlvarez Montaner, Characteristic cycles of local cohomology modules of monomial ideals. J. Pure Appl. Algebra **150**, 1–25 (2000)
4. J. Àlvarez Montaner, Characteristic cycles of local cohomology modules of monomial ideals II. J. Pure Appl. Algebra **192**, 1–20 (2004)
5. J. Àlvarez Montaner, Some numerical invariants of local rings. Proc. Am. Math. Soc. **132**, 981–986 (2004)
6. J. Àlvarez Montaner, Operations with regular holonomic D-modules with support a normal crossing. J. Symb. Comput. **40**, 999–1012 (2005)
7. J. Àlvarez Montaner, R. García López, S. Zarzuela, Local cohomology, arrangements of subspaces and monomial ideals. Adv. Math. **174**, 35–56 (2003)
8. J. Àlvarez Montaner, S. Zarzuela, Linearization of local cohomology modules, in *Commutative Algebra: Interactions with Algebraic Geometry*, ed. by L.L. Avramov, M. Chardin, M. Morales, C. Polini. Contemporary Mathematics, vol. 331 (American Mathematical Society, Providence, 2003), pp. 1–11
9. J. Àlvarez Montaner, A. Leykin, Computing the support of local cohomology modules. J. Symb. Comput. **41**, 1328–1344 (2006)
10. J. Àlvarez Montaner, A. Vahidi, Lyubeznik numbers of monomial ideals. Trans. Am. Math. Soc. (2011) [arXiv/1107.5230] (to appear)
11. I. Anwar, D. Popescu, Stanley Conjecture in small embedding dimension. J. Algebra **318**, 1027–1031 (2007)
12. J. Apel, On a conjecture of R. P. Stanley. Part I-Monomial Ideals. J. Algebr. Comb. **17**, 36–59 (2003)
13. J. Apel, On a conjecture of R. P. Stanley. Part II-Quotients modulo monomial ideals. J. Algebr. Comb. **17**, 57–74 (2003)
14. S. Bandari, K. Divaani-Aazar, A. Soleyman Jahan, Filter-regular sequences, almost complete intersections and Stanley's conjecture (2011) [arXiv:1112.5159]
15. F. Barkats, Calcul effectif de groupes de cohomologie locale à support dans des idéaux monomiaux. Ph.D. Thesis, Univ. Nice-Sophia Antipolis, 1995
16. D. Bayer, I. Peeva, B. Sturmfels, Monomial resolutions. Math. Res. Lett. **5**, 31–46 (1998)
17. D. Bayer, B. Sturmfels, Cellular resolutions of monomial modules. J. Reine Angew. Math. **502**, 123–140 (1998)

A.M. Bigatti et al. (eds.), *Monomial Ideals, Computations and Applications*,
Lecture Notes in Mathematics 2083, DOI 10.1007/978-3-642-38742-5,
© Springer-Verlag Berlin Heidelberg 2013

18. C. Biró, D. Howard, M. Keller, W. Trotter, S. Young, Interval partitions and Stanley depth. J. Comb. Theory Ser. A **117**, 475–482 (2010)
19. I. Bermejo, P. Gimenez, Saturation and Castelnuovo-Mumford regularity. J. Algebra **303**, 592–617 (2006)
20. J. Biermann, Cellular structure on the minimal resolution of the edge ideal of the complement of the n-cycle (2011). Preprint
21. J.E. Björk, *Rings of Differential Operators* (North Holland Mathematics Library, Amsterdam, 1979)
22. A. Björner, M.L. Wachs, Shellable nonpure complexes and posets II. Trans. Am. Math. Soc. **349**, 3945–3975 (1997)
23. M. Blickle, Lyubeznik's numbers for cohomologically isolated singularities. J. Algebra **308**, 118–123 (2007)
24. M. Blickle, R. Bondu, Local cohomology multiplicities in terms of étale cohomology. Ann. Inst. Fourier **55**, 2239–2256 (2005)
25. R. Boldini, Critical cones of characteristic varieties. Trans. Am. Math. Soc. **365**, 143–160 (2013)
26. A. Borel, in *Algebraic D-Modules*. Perspectives in Mathematics (Academic, London, 1987)
27. R. Bouchat, H.T. Hà, A. O'Keefe, Path ideals of rooted trees and their graded Betti numbers. J. Comb. Theory Ser. A **118**, 2411–2425 (2011)
28. M. Brodmann, The asymptotic nature of the analytic spread. Math. Proc. Camb. Philos. Soc. **86**, 35–39 (1979)
29. M.P. Brodmann, R.Y. Sharp, in *Local Cohomology, An Algebraic Introduction with Geometric Applications*. Cambridge Studies in Advanced Mathematics, vol. 60 (Cambridge University Press, Cambridge, 1998)
30. W. Bruns, J. Herzog, *Cohen-Macaulay rings*, revised edn. (Cambridge University Press, Cambridge, 1998)
31. W. Bruns, C. Krattenthaler, J. Uliczka, Stanley decompositions and Hilbert depth in the Koszul complex. J. Commut. Algebra **2**, 327–357 (2010)
32. W. Bruns, C. Krattenthaler, J. Uliczka, Hilbert depth of powers of the maximal ideal, in *Commutative Algebra and Its Connections to Geometry (PASI 2009)*, ed. by A. Corso, C. Polini. Contemporary Mathematics, vol. 555 (American Mathematical Society, Providence, 2011), pp. 1–12
33. J. Chen, S. Morey, A. Sung, The stable set of associated primes of the ideal of a graph. Rocky Mt. J. Math. **32**, 71–89 (2002)
34. R.-X. Chen, Minimal free resolutions of linear edge ideals. J. Algebra **324**, 3591–3613 (2010)
35. M. Chudnovsky, N. Robertson, P. Seymour, R. Thomas, The strong perfect graph theorem. Ann. Math. (2) **164**, 51–229 (2006)
36. M. Cimpoeaş, Some remarks on the Stanley depth for multigraded modules. Le Matematiche **LXIII**, 165–175 (2008)
37. M. Cimpoeaş, Stanley depth of monomial ideals with small number of generators. Cent. Eur. J. Math. **7**, 629–634 (2009)
38. M. Cimpoeaş, Stanley depth of squarefree Veronese ideals (2009). arXiv:0907.1232 [math.AC]
39. CoCoATeam, CoCoA: A system for doing Computations in Commutative Algebra (2011). Available at http://cocoa.dima.unige.it
40. A. Conca, E. De Negri, M-Sequences, graph ideals and ladder ideals of linear type. J. Algebra **211**, 599–624 (1999)
41. A. Conca, J. Herzog, Castelnuovo-Mumford regularity of products of ideals. Collect. Math. **54**, 137–152 (2003)
42. D.W. Cook II, Simplicial decomposability. J. Softw. Algebra Geom. **2**, 20–23 (2010)
43. A. Corso, U. Nagel, Specializations of Ferrers ideals. J. Algebr. Comb. **28**, 425–437 (2008)
44. A. Corso, U. Nagel, Monomial and toric ideals associated to Ferrers graphs. Trans. Am. Math. Soc. **361**, 1371–1395 (2009)

45. S.C. Coutinho, in *A Primer of Algebraic 𝒟-Modules*. London Mathematical Society Student Texts (Cambridge University Press, Cambridge, 1995)
46. K. Dalili, M. Kummini, Dependence of Betti numbers on characteristic (2010) [arXiv:1009.4243]
47. W. Decker, G.-M. Greuel, G. Pfister, H. Schönemann, SINGULAR 3-1-3 — A computer algebra system for polynomial computations (2011). http://www.singular.uni-kl.de
48. A. Dochtermann, A. Engström, Algebraic properties of edge ideals via combinatorial topology. Electron. J. Combin. **16** (2009). Special volume in honor of Anders Bjorner, Research Paper 2, 24 pp.
49. A. Dochtermann, A. Engström, Cellular resolutions of cointerval ideals. Math. Z. **270**, 145–163 (2012)
50. A. Dress, A new algebraic criterion for shellability. Beitrage zur Alg. und Geom. **34**, 45–55 (1993)
51. J.A. Eagon, V. Reiner, Resolutions of Stanley-Reisner rings and Alexander duality. J. Pure Appl. Algebra **130**, 265–275 (1998)
52. D. Eisenbud, in *Commutative Algebra with a View Towards Algebraic Geometry*. GTM, vol. 150 (Springer, Berlin, 1995)
53. D. Eisenbud, G. Fløystad, F.O. Schreyer, Sheaf cohomology and free resolutions over exterior algebras. Trans. Am. Math. Soc. **355**, 4397–4426 (2003)
54. D. Eisenbud, M. Green, K. Hulek, S. Popescu, Restricting linear syzygies: Algebra and geometry. Compos. Math. **141**, 1460–1478 (2005)
55. D. Eisenbud, M. Mustață, M. Stillman, Cohomology on toric varieties and local cohomology with monomial supports. J. Symb. Comput. **29**, 583–600 (2000)
56. S. Eliahou, M. Kervaire, Minimal resolutions of some monomial ideals. J. Algebra **129**, 1–25 (1990)
57. E. Emtander, A class of hypergraphs that generalizes chordal graphs. Math. Scand. **106**, 50–66 (2010)
58. E. Emtander, Betti numbers of hypergraphs. Commun. Algebra **37**, 1545–1571 (2009)
59. S. Faridi, The facet ideal of a simplicial complex. Manuscripta Math. **109**, 159–174 (2002)
60. G. Fatabbi, On the resolution of ideals of fat points. J. Algebra **242**, 92–108 (2001)
61. O. Fernández-Ramos, P. Gimenez, First nonlinear syzygies of ideals associated to graphs. Commun. Algebra **37**, 1921–1933 (2009)
62. G. Fløystad, J. Herzog, Grö;bner bases of syzygies and Stanley depth. J. Algebra **328**, 178–189 (2011)
63. H.B. Foxby, Bounded complexes of flat modules. J. Pure Appl. Algebra **15**, 149–172 (1979)
64. C.A. Francisco, Resolutions of small sets of fat points. J. Pure Appl. Algebra **203**, 220–236 (2005)
65. C. Francisco, H.T. Hà, A. Van Tuyl, Splittings of monomial ideals. Proc. Am. Math. Soc. **137**, 3271–3282 (2009)
66. C.A. Francisco, H.T. Hà, A. Van Tuyl, A conjecture on critical graphs and connections to the persistence of associated primes. Discrete Math. **310**, 2176–2182 (2010)
67. C.A. Francisco, H.T. Hà, A. Van Tuyl, Associated primes of monomial ideals and odd holes in graphs. J. Algebr. Comb. **32**, 287–301 (2010)
68. C.A. Francisco, H.T. Hà, A. Van Tuyl, Colorings of hypergraphs, perfect graphs, and associated primes of powers of monomial ideals. J. Algebra **331**, 224–242 (2011)
69. C.A. Francisco, A. Hoefel, A. Van Tuyl, EdgeIdeals: A package for (hyper)graphs. J. Softw. Algebra Geom. **1**, 1–4 (2009)
70. R. Fröberg, On Stanley-Reisner rings, in *Topics in Algebra, Part 2 (Warsaw, 1988)*. PWN, vol. 26 (Banach Center Publ., Warsaw, 1990), pp. 57–70
71. O. Gabber, The integrability of the Characteristic Variety. Am. J. Math. **103**, 445–468 (1981)
72. A. Galligo, M. Granger, Ph. Maisonobe, D-modules et faisceaux pervers dont le support singulier est un croisement normal. Ann. Inst. Fourier **35**, 1–48 (1985)

73. A. Galligo, M. Granger, Ph. Maisonobe, D-modules et faisceaux pervers dont le support singulier est un croisement normal. II, in *Differential Systems and Singularities (Luminy, 1983)*. Astérisque **130**, 240–259 (1985)

74. R. Garcia, C. Sabbah, Topological computation of local cohomology multiplicities. Collect. Math. **49**, 317–324 (1998)

75. V. Gasharov, I. Peeva, V. Welker, The lcm-lattice in monomial resolutions. Math. Res. Lett. **6**, 521–532 (1999)

76. M. Ge, J. Lin, Y.H. Shen, On a conjecture of Stanley depth of squarefree Veronese ideals. Commun. Algebra **40**, 2720–2731 (2012)

77. M. Goresky, R. MacPherson, in *Stratified Morse Theory*. Ergebnisse Series, vol. 14 (Springer, Berlin, 1988)

78. S. Goto, K. Watanabe, On Graded Rings, II ($\mathbb{Z}^n$- graded rings). Tokyo J. Math. **1**, 237–261 (1978)

79. H.G. Gräbe, The canonical module of a Stanley-Reisner ring. J. Algebra **86**, 272–281 (1984)

80. D. Grayson, M. Stillman, Macaulay2, A software system for research in algebraic geometry (2011). Available at: http://www.math.uiuc.edu/Macaulay2

81. A. Grothendieck, in *Local Cohomology*. Lecture Notes in Mathematics, vol. 41 (Springer, Berlin, 1967)

82. A. Grothendieck, J. Dieudonné, *Éléments de géométrie algébrique IV. Étude locale des schémas et des morphismes de schémas* Inst. Hautes Études Sci. Publ. Math. **28** (1966)

83. H.T. Hà, A. Van Tuyl, Splittable ideals and the resolution of monomial ideals. J. Algebra **309**, 405–425 (2007)

84. H.T. Hà, A. Van Tuyl, Resolutions of square-free monomial ideals via facet ideals: A survey. Contemp. Math. **448**, 91–117 (2007)

85. H.T. Hà, A. Van Tuyl, Monomial ideals, edge ideals of hypergraphs, and their graded Betti numbers. J. Algebr. Combin. **27**, 215–245 (2008)

86. J. He, A. Van Tuyl, Algebraic properties of the path ideal of a tree. Commun. Algebra **38**, 1725–1742 (2010)

87. M. Hellus, A note on the injective dimension of local cohomology modules. Proc. Am. Math. Soc. **136**, 2313–2321 (2008)

88. J. Herzog, T. Hibi, Componentwise linear ideals. Nagoya Math. J. **153**, 141–153 (1999)

89. J. Herzog, T. Hibi, The depth of powers of an ideal. J. Algebra **291**, 534–550 (2005)

90. J. Herzog, T. Hibi, Distributive lattices, bipartite graphs and Alexander duality. J. Algebr. Comb. **22**, 289–302 (2005)

91. J. Herzog, T. Hibi, in *Monomial Ideals*. GTM, vol. 260 (Springer, Berlin, 2010)

92. J. Herzog, T. Hibi, F. Hreinsdóttir, T. Kahle, J. Rauh, Binomial edge ideals and conditional independence statements. Adv. Appl. Math. **45**, 317–333 (2010)

93. J. Herzog, T. Hibi, X. Zheng, Monomial ideals whose powers have a linear resolution. Math. Scand. **95**, 23–32 (2004)

94. J. Herzog, S. Iyengar, Koszul modules. J. Pure Appl. Algebra **201**, 154–188 (2005)

95. J. Herzog, D. Popescu, Finite filtrations of modules and shellable multicomplexes. Manuscripta Math. **121**, 385–410 (2006)

96. J. Herzog, D. Popescu, M. Vladoiu, On the Ext-modules of ideals of Borel type. Contemp. Math. **331**, 171–186 (2003)

97. J. Herzog, D. Popescu, M. Vladoiu, Stanley depth and size of a monomial ideal. Proc. Am. Math. Soc. **140**, 493–504 (2012)

98. J. Herzog, A. Soleyman Jahan, S. Yassemi, Stanley decompositions and partitionable simplicial complexes. J. Algebr. Comb. **27**, 113–125 (2008)

99. J. Herzog, A. Soleyman Jahan, X. Zheng, Skeletons of monomial ideals. Math. Nachr. **283**, 1403–1408 (2010)

100. J. Herzog, H. Takayama, Resolutions by mapping cones. The Roos Festschrift vol. 2. Homology Homotopy Appl. **4**, 277–294 (2002)

101. J. Herzog, M. Vladoiu, X. Zheng, How to compute the Stanley depth of a monomial ideal (2007) [arXiv:0712.2308v1]

102. J. Herzog, M. Vladoiu, X. Zheng, How to compute the Stanley depth of a monomial ideal. J. Algebra **322**, 3151–3169 (2009)
103. T. Hibi, Distributive lattices, affine semigroup rings and algebras with straightening laws, in *Commutative Algebra and Combinatorics*, ed. by M. Nagata, H. Matsumura. Advanced Studies in Pure Mathematics, vol. 11 (North-Holland, Amsterdam, 1987), pp. 93–109
104. T. Hibi, Quotient algebras of Stanley-Reisner rings and local cohomology. J. Algebra **140**, 336–343 (1991)
105. T. Hibi, K. Kimura, S. Murai, Betti numbers of chordal graphs and f-vectors of simplicial complexes. J. Algebra **323**, 1678–1689 (2010)
106. A. Hoefel, G. Whieldon, Linear quotients of the square of the edge ideal of the anticycle (2011). Preprint [arXiv:1106.2348v2]
107. N. Horwitz, Linear resolutions of quadratic monomial ideals. J. Algebra **318**, 981–1001 (2007)
108. S. Ho sten, G.G. Smith, Monomial ideals, in *Computations in Algebraic Geometry with Macaulay 2*. Algorithms and Computations in Mathematics, vol. 8 (Springer, New York, 2001), pp. 73–100
109. C. Huneke, Problems on local cohomology modules, in *Free Resolutions in Commutative Algebra and Algebraic Geometry (Sundance, 1990)*. Research Notes in Mathematics, vol. 2 (Jones and Barthlett Publishers, Boston, 1994), pp. 93–108
110. C. Huneke, R.Y. Sharp, Bass numbers of local cohomology modules. Trans. Am. Math. Soc. **339**, 765–779 (1993)
111. M. Ishaq, Upper bounds for the Stanley depth. Commun. Algebra **40**, 87–97 (2012)
112. M. Ishaq, Values and bounds of the Staney depth. Carpathian J. Math. **27**, 217–224 (2011) [arXiv:1010.4692]
113. S. Iyengar, G. Leuschke, A. Leykin, C. Miller, E. Miller, A. Singh, U. Walther, in *Twenty-Four Hours of Local Cohomology*. Graduate Studies in Mathematics, vol. 87 (American Mathematical Society, Providence, 2007)
114. S. Jacques, Betti Numbers of Graph Ideals. Ph.D. Thesis, University of Sheffield, 2004 [arXiv:math/0410107v1]
115. M. Janet, Les modules des formes algébriques et la théorie générale des systèmes différentiels. Ann. Sci. École Norm. Sup. **41**, 27–65 (1924)
116. M. Jöllenbeck, V. Welker, Minimal resolutions via algebraic discrete Morse theory. Mem. Am. Math. Soc. **197** vi+74 pp. (2009)
117. T. Kaiser, M. Stehlik, R. Skrekovski, Replication in critical graphs and the persistence of monomial ideals (2013). Preprint [arXiv:1301.6983.v2]
118. M. Katzman, Characteristic-independence of Betti numbers of graph ideals. J. Comb. Theory Ser. A **113**, 435–454 (2006)
119. M.T. Keller, Y-H. Shen, N. Streib, S.T. Young, On the Stanley depth of squarefree Veronese ideals. J. Algebr. Comb. **33**, 313–324 (2011)
120. M.T. Keller, S.J. Young, Stanley depth of squarefree monomial ideals. J. Algebra **322**, 3789–3792 (2009)
121. K. Kimura, Non-vanishingness of Betti numbers of edge ideals (2011). Preprint [arXiv:1110.2333v3]
122. F. Kirwan, in *An Introduction to Intersection Homology Theory*. Pitman Research Notes in Mathematics (Longman Scientific and Technical, Harlow, 1988); copublished in the United States with Wiley, New York
123. S. Khoroshkin, D-modules over arrangements of hyperplanes. Commun. Algebra **23**, 3481–3504 (1995)
124. S. Khoroshkin, A. Varchenko, Quiver D-modules and homology of local systems over arrangements of hyperplanes. Inter. Math. Res. Papers, 2006
125. M. Kummini, Regularity, depth and arithmetic rank of bipartite edge ideals. J. Algebr. Comb. **30**, 429–445 (2009)
126. A. Leykin, D-modules for Macaulay 2, in *Mathematical Software: ICMS 2002* (World Scientific, Singapore, 2002), pp. 169–179

127. A. Leykin, M. Stillman, H. Tsai, D-modules for Macaulay 2 (2011). Available at: http://people.math.gatech.edu/~aleykin3/Dmodules
128. G. Lyubeznik, On the local cohomology modules $H^i_{\mathfrak{A}}(R)$ for ideals $\mathfrak{A}$ generated by monomials in an R-sequence, in *Complete Intersections*, ed. by S. Greco, R. Strano. Lecture Notes in Mathematics, vol. 1092 (Springer, Berlin, 1984), pp. 214–220
129. G. Lyubeznik, On the arithmetical rank of monomial ideals. J. Algebra **112**, 86–89 (1988)
130. G. Lyubeznik, A new explicit finite free resolution of ideals generated by monomials in an R-sequence. J. Pure Appl. Algebra **51**, 193–195 (1988)
131. G. Lyubeznik, The minimal non-Cohen-Macaulay monomial ideals. J. Pure Appl. Algebra **51**, 261–266 (1988)
132. G. Lyubeznik, Finiteness properties of local cohomology modules (an application of D-modules to commutative algebra). Invent. Math. **113**, 41–55 (1993)
133. G. Lyubeznik, F-modules: Applications to local cohomology and D-modules in characteristic $p > 0$. J. Reine Angew. Math. **491**, 65–130 (1997)
134. G. Lyubeznik, A partial survey of local cohomology modules, in *Local Cohomology Modules and Its Applications (Guanajuato, 1999)*. Lecture Notes in Pure and Applied Mathematics, vol. 226 (Marcel Dekker, New York, 2002), pp. 121–154
135. G. Lyubeznik, On some local cohomology modules. Adv. Math. **213**, 621–643 (2007)
136. D. Maclagan, G.G. Smith, Smooth and irreducible multigraded Hilbert schemes. Adv. Math. **223**, 1608–1631 (2010)
137. B. Malgrange, L'involutivité des caractéristiques des systèmes différentiels et microdifférentiels, in *Lecture Notes in Mathematics*, vol. 710 (Springer, Berlin, 1979), pp. 277–289
138. J. Martinez-Bernal, S. Morey, R. Villarreal, Associated primes of powers of edge ideals (2011). Preprint [arXiv:1103.0992v3]
139. Z. Mebkhout, in *Le formalisme des six opérations de Grothendieck pour les $\mathscr{D}_X$-modules cohérents*. Travaux en Cours, vol. 35 (Hermann, Paris, 1989)
140. E. Miller, The Alexander duality functors and local duality with monomial support. J. Algebra **231**, 180–234 (2000)
141. E. Miller, B. Sturmfels, in *Combinatorial Commutative Algebra*. Graduate Texts in Mathematics, vol. 227 (Springer, New York, 2005)
142. E. Miller, B. Sturmfels, K. Yanagawa, Generic and cogeneric monomial ideals. J. Symb. Comput. **29**, 691–708 (2000)
143. S. Moradi, D. Kiani, Bounds for the regularity of edge ideal of vertex decomposable and shellable graphs. Bull. Iran. Math. Soc. **36**, 267–277 (2010)
144. S. Morey, E. Reyes, R. Villarreal, Cohen-Macaulay, shellable and unmixed clutters with a perfect matching of König type. J. Pure Appl. Algebra **212**, 1770–1786 (2008)
145. S. Morey, R. Villarreal, Edge ideals: Algebraic and combinatorial properties, in *Progress in Commutative Algebra*, vol. 1 (de Gruyter, Berlin, 2012), pp. 85–126
146. M. Mustaţă, Local Cohomology at Monomial Ideals. J. Symb. Comput. **29**, 709–720 (2000)
147. M. Mustaţă, Vanishing theorems on toric varieties. Tohoku Math. J. **54**, 451–470 (2002)
148. S. Nasir, Stanley decompositions and localization. Bull. Math. Soc. Sci. Math. Roumanie **51**, 151–158 (2008)
149. S. Nasir, A. Rauf, Stanley decompositions in localized polynomial rings. Manuscripta Math. **135**, 151–164 (2011)
150. E. Nevo, Regularity of edge ideals of C_4-free graphs via the topology of the lcm-lattice. J. Comb. Theory Ser. A **118**, 491–501 (2011)
151. E. Nevo, I. Peeva, Linear resolutions of powers of edge ideals (2010). Preprint
152. T. Oaku, Algorithms for b-functions, restrictions and local cohomology groups of D-modules. Adv. Appl. Math. **19**, 61–105 (1997)
153. T. Oaku, N. Takayama, Algorithms for D-modules—restriction, tensor product, localization, and local cohomology groups. J. Pure Appl. Algebra **156**, 267–308 (2001)
154. R. Okazaki, A lower bound of Stanley depth of monomial ideals. J. Commut. Algebra **3**, 83–88 (2011)

155. R. Okazaki, K. Yanagawa, Alexander duality and Stanley depth of multigraded modules. J. Algebra **340**, 35–52 (2011)
156. I. Peeva, *Graded Syzygies* (Springer, New York, 2010)
157. I. Peeva, M. Velasco, Frames and degenerations of monomial resolutions. Trans. Am. Math. Soc. **363**, 2029–2046 (2011)
158. F. Pham, in *Singularités des systèmes différentiels de Gauss-Manin*. Progress in Mathematics, vol. 2 (Birkhäuser, Basel, 1979)
159. A. Popescu, Special Stanley Decompositions. Bull. Math. Soc. Sci. Math. Roumanie **53**, 361–372 (2010)
160. D. Popescu, Stanley conjecture on intersections of four monomial prime ideals (2010) [arXiv. AC/1009.5646]
161. D. Popescu, Stanley depth of multigraded modules. J. Algebra **321**, 2782–2797 (2009)
162. D. Popescu, M.I. Qureshi, Computing the Stanley depth. J. Algebra **323**, 2943–2959 (2010)
163. M.R. Pournaki, S.A. Seyed Fakhari, S. Yassemi, Stanley depth of powers of the edge ideal of a forest. Proc. Am. Math Soc. (Electronically published on June 7, 2013)
164. A. Rauf, Depth and Stanley depth of multigraded modules. Commun. Algebra **38**, 773–784 (2010)
165. G. Rinaldo, A routine to compute the Stanley depth of I/J (2011). http://ww2.unime.it/algebra/rinaldo/sdepth.html
166. T. Römer, Generalized Alexander duality and applications. Osaka J. Math. **38**, 469–485 (2001)
167. T. Römer, On minimal graded free resolutions. Ph.D. Thesis, Essen, 2001
168. M. Sato, T. Kawai, M. Kashiwara, Microfunctions and pseudo-differential equations, in *Lecture Notes in Mathematics*, vol. 287 (Springer, New York, 1973), pp. 265–529
169. E. Scheinerman, D. Ullman, *Fractional Graph Theory. A Rational Approach to the Theory of Graphs* (Wiley, New York, 1997)
170. P. Schenzel, On Lyubeznik's invariants and endomorphisms of local cohomology modules. J. Algebra **344**, 229–245 (2011)
171. S.A. Seyed Fakhari, Stanley depth of the integral closure of monomial ideals (2012) [arXiv: 1205.6971]
172. S.A. Seyed Fakhari, Stanley depth of weakly polymatroidal ideals and squarefree monomial ideals (2013) [arXiv:1302.5837]
173. R.Y. Sharp, *Steps in Commutative Algebra*, 2nd edn. (Cambridge University Press, Cambridge, 2000)
174. Y-H. Shen, Stanley depth of complete intersection monomial ideals and upper-discrete partitions. J. Algebra **321**, 1285–1292 (2009)
175. A. Simis, W. Vasconcelos, R.H. Villarreal, On the ideal theory of graphs. J. Algebra **167**, 389–416 (1994)
176. A.M. Simon, Some homological properties of complete modules. Math. Proc. Camb. Philos. Soc. **108**, 231–246 (1990)
177. G.G. Smith, Irreducible components of characteristic varieties. J. Pure Appl. Algebra **165**, 291–306 (2001)
178. A. Soleyman Jahan, Prime filtrations of monomial ideals and polarizations. J. Algebra **312**, 1011–1032 (2007)
179. A. Soleyman Jahan, Prime filtrations and Stanley decompositions of squarefree modules and Alexander duality. Manuscripta Math. **130**, 533–550 (2009)
180. R. Stanley, The Upper Bound Conjecture and Cohen-Macaulay rings. Stud. Appl. Math. **54**, 135–142 (1975)
181. R.P. Stanley, Linear Diophantine equations and local cohomology. Invent. Math. **68**, 175–193 (1982)
182. R.P. Stanley, *Combinatorics and Commutative Algebra* (Birkhäuser, Basel, 1983)
183. R.P. Stanley, Positivity Problems and Conjectures in Algebraic Combinatorics, in *Mathematics: Frontiers and Perspectives*, ed. by V. Arnold, M. Atiyah, P. Lax, B. Mazur (American Mathematical Society, Providence, 2000), pp. 295–319

184. B. Sturmfels, The co-Scarf resolution, in *Commutative Algebra, Algebraic Geometry, and Computational Methods*, ed. by D. Eisenbud (Springer, Singapore, 1999), pp. 315–320

185. B. Sturmfels, S. Sullivant, Combinatorial secant varieties. Pure Appl. Math. Q. **2**(3), Part 1, 867–891 (2006)

186. B. Sturmfels, N. White, Computing combinatorial decompositions of rings. Combinatorica **11**, 275–293 (1991)

187. N. Terai, Local cohomology modules with respect to monomial ideals (1999). Preprint

188. D. Taylor, Ideals genreated by monomials in an *R*-sequence. Ph.D. Thesis, University of Chicago, 1961

189. G. Valla, Betti numbers of some monomial ideals. Proc. Am. Math. Soc. **133**, 57–63 (2005)

190. A. Van Tuyl, Sequentially Cohen-Macaulay bipartite graphs: vertex decomposability and regularity. Arch. Math. (Basel) **93**, 451–459 (2009)

191. M. Velasco, Minimal free resolutions that are not supported by a CW-complex. J. Algebra **319**, 102–114 (2008)

192. R.H. Villarreal, Cohen-Macaulay graphs. Manuscripta Math. **66**, 277–293 (1990)

193. R.H. Villarreal, in *Monomial Algebras*. Monographs and Textbooks in Pure and Applied Mathematics, vol. 238 (Marcel Dekker, New York, 2001)

194. U. Walther, Algorithmic computation of local cohomology modules and the cohomological dimension of algebraic varieties. J. Pure Appl. Algebra **139**, 303–321 (1999)

195. U. Walther, D-modules and cohomology of varieties, in *Computations in Algebraic Geometry with Macaulay 2*. Algorithms and Computations in Mathematics, vol. 8 (Springer, Berlin, 2002), pp. 281–323

196. G. Whieldon, Jump sequences of edge ideals (2010). Preprint [arXiv:1012.0108v1]

197. R. Woodroofe, Matchings, coverings, and Castelnuovo-Mumford regularity (2010). Preprint [arXiv:1009.2756]

198. R. Woodroofe, Chordal and sequentially Cohen-Macaulay clutters. Electron. J. Comb. **18**, Paper 208 (2011)

199. K. Yanagawa, Alexander duality for Stanley-Reisner rings and squarefree ℕ-graded modules. J. Algebra **225**, 630–645 (2000)

200. K. Yanagawa, Bass numbers of local cohomology modules with supports in monomial ideals. Math. Proc. Camb. Philos. Soc. **131**, 45–60 (2001)

201. K. Yanagawa, Sheaves on finite posets and modules over normal semigroup rings. J. Pure Appl. Algebra **161**, 341–366 (2001)

202. K. Yanagawa, Derived category of squarefree modules and local cohomology with monomial ideal support. J. Math. Soc. Jpn. **56**, 289–308 (2004)

203. X. Zheng, Resolutions of facet ideals. Commun. Algebra **32**, 2301–2324 (2004)

LECTURE NOTES IN MATHEMATICS

 Springer

Edited by J.-M. Morel, B. Teissier; P.K. Maini

Editorial Policy (for Multi-Author Publications: Summer Schools / Intensive Courses)

1. Lecture Notes aim to report new developments in all areas of mathematics and their applications - quickly, informally and at a high level. Mathematical texts analysing new developments in modelling and numerical simulation are welcome. Manuscripts should be reasonably selfcontained and rounded off. Thus they may, and often will, present not only results of the author but also related work by other people. They should provide sufficient motivation, examples and applications. There should also be an introduction making the text comprehensible to a wider audience. This clearly distinguishes Lecture Notes from journal articles or technical reports which normally are very concise. Articles intended for a journal but too long to be accepted by most journals, usually do not have this "lecture notes" character.

2. In general SUMMER SCHOOLS and other similar INTENSIVE COURSES are held to present mathematical topics that are close to the frontiers of recent research to an audience at the beginning or intermediate graduate level, who may want to continue with this area of work, for a thesis or later. This makes demands on the didactic aspects of the presentation. Because the subjects of such schools are advanced, there often exists no textbook, and so ideally, the publication resulting from such a school could be a first approximation to such a textbook. Usually several authors are involved in the writing, so it is not always simple to obtain a unified approach to the presentation.

   For prospective publication in LNM, the resulting manuscript should not be just a collection of course notes, each of which has been developed by an individual author with little or no coordination with the others, and with little or no common concept. The subject matter should dictate the structure of the book, and the authorship of each part or chapter should take secondary importance. Of course the choice of authors is crucial to the quality of the material at the school and in the book, and the intention here is not to belittle their impact, but simply to say that the book should be planned to be written by these authors jointly, and not just assembled as a result of what these authors happen to submit.

   This represents considerable preparatory work (as it is imperative to ensure that the authors know these criteria before they invest work on a manuscript), and also considerable editing work afterwards, to get the book into final shape. Still it is the form that holds the most promise of a successful book that will be used by its intended audience, rather than yet another volume of proceedings for the library shelf.

3. Manuscripts should be submitted either online at www.editorialmanager.com/lnm/ to Springer's mathematics editorial, or to one of the series editors. Volume editors are expected to arrange for the refereeing, to the usual scientific standards, of the individual contributions. If the resulting reports can be forwarded to us (series editors or Springer) this is very helpful. If no reports are forwarded or if other questions remain unclear in respect of homogeneity etc, the series editors may wish to consult external referees for an overall evaluation of the volume. A final decision to publish can be made only on the basis of the complete manuscript; however a preliminary decision can be based on a pre-final or incomplete manuscript. The strict minimum amount of material that will be considered should include a detailed outline describing the planned contents of each chapter.

   Volume editors and authors should be aware that incomplete or insufficiently close to final manuscripts almost always result in longer evaluation times. They should also be aware that parallel submission of their manuscript to another publisher while under consideration for LNM will in general lead to immediate rejection.

4. Manuscripts should in general be submitted in English. Final manuscripts should contain at least 100 pages of mathematical text and should always include

   - a general table of contents;
   - an informative introduction, with adequate motivation and perhaps some historical remarks: it should be accessible to a reader not intimately familiar with the topic treated;
   - a global subject index: as a rule this is genuinely helpful for the reader.

   Lecture Notes volumes are, as a rule, printed digitally from the authors' files. We strongly recommend that all contributions in a volume be written in the same LaTeX version, preferably LaTeX2e. To ensure best results, authors are asked to use the LaTeX2e style files available from Springer's web-server at
   ftp://ftp.springer.de/pub/tex/latex/svmonot1/ (for monographs) and
   ftp://ftp.springer.de/pub/tex/latex/svmultt1/ (for summer schools/tutorials).
   Additional technical instructions, if necessary, are available on request from:
   lnm@springer.com.

5. Careful preparation of the manuscripts will help keep production time short besides ensuring satisfactory appearance of the finished book in print and online. After acceptance of the manuscript authors will be asked to prepare the final LaTeX source files and also the corresponding dvi-, pdf- or zipped ps-file. The LaTeX source files are essential for producing the full-text online version of the book. For the existing online volumes of LNM see:
   http://www.springerlink.com/openurl.asp?genre=journal&issn=0075-8434.
   The actual production of a Lecture Notes volume takes approximately 12 weeks.

6. Volume editors receive a total of 50 free copies of their volume to be shared with the authors, but no royalties. They and the authors are entitled to a discount of 33.3 % on the price of Springer books purchased for their personal use, if ordering directly from Springer.

7. Commitment to publish is made by letter of intent rather than by signing a formal contract. Springer-Verlag secures the copyright for each volume. Authors are free to reuse material contained in their LNM volumes in later publications: a brief written (or e-mail) request for formal permission is sufficient.

Addresses:
Professor J.-M. Morel, CMLA,
École Normale Supérieure de Cachan,
61 Avenue du Président Wilson, 94235 Cachan Cedex, France
E-mail: morel@cmla.ens-cachan.fr

Professor B. Teissier, Institut Mathématique de Jussieu,
UMR 7586 du CNRS, Équipe "Géométrie et Dynamique",
175 rue du Chevaleret,
75013 Paris, France
E-mail: teissier@math.jussieu.fr

For the "Mathematical Biosciences Subseries" of LNM:

Professor P. K. Maini, Center for Mathematical Biology,
Mathematical Institute, 24-29 St Giles,
Oxford OX1 3LP, UK
E-mail : maini@maths.ox.ac.uk

Springer, Mathematics Editorial I,
Tiergartenstr. 17,
69121 Heidelberg, Germany,
Tel.: +49 (6221) 4876-8259
Fax: +49 (6221) 4876-8259
E-mail: lnm@springer.com